Trolling Euclid

An Irreverent Guide to Nine of Mathematics' Most Important Problems

Tom Wright

Contents

INTRODUCTION   1

1   Welcome to My Book!   1

2   Verbiage   9

3   Odds and Ends   11

THE RIEMANN HYPOTHESIS   13

4   Historical Background: Straight Cash, Homey and Other Mathematical Concepts   14

5   Make My Funk a Z-Func(tion)   18

6   The Zeta Function: Magical, Mystical, and....Dear God, What Is That Thing?   22

7   Application: Primes on Parade   28

8   Appendix A: Analytic Continuation   33

THE GENERALIZED RIEMANN HYPOTHESIS   37

9   Generalized Riemann Hypothesis: Because The Riemann Hypothesis Wasn't Hard Enough   38

CONTENTS   iii

10 Let's Get Ready to Rumble and Rewrite the Riemann Hypothesis!   46

11 GRH Goes To The Races: Daddy Needs a New Pair of Shoes!   51

12 Other Corollaries of GRH   59

THE ABC CONJECTURE   67

13 ABC: What the Alphabet Looks Like When D Through Z are Eliminated   68

14 Radical! Mason and Oesterle's Excellent Adventure   72

15 Towards A Meaningful ABC   75

16 Current State Of Affairs: Where We Are and Things That The ABC Conjecture Would Prove   81

17 Appendix B: Why ABC gives us Fermat's Last Theorem   85

THE BIRCH-SWINNERTON-DYER CONJECTURE   89

18 Preface   90

19 Elliptic Curves: Nothing to Do With Ellipses   92

20 Modular Arithmetic: Why Telling Time Actually Counts as Doing Math   97

21 L-Functions: Convoluted Functions with Weird Powers   103

22 BSD: Carrying the One   108

CONTENTS   iv

23 Digression: Equations and Diagrams That Are Required to
   Go in Any Write-up of BSD   111

BSD II: THE PROBLEM STRIKES BACK   113

24 More BSD: The Stronger, Better, Faster Version   114

25 Elliptic Curve Structure: Like Regular Addition But With
   Way More Symbols   117

26 How Zero is Zero?   124

27 Finally: The Full, All-Powerful, Earth-Shattering, Cavity-
   Reducing, Baldness-Curing Birch-Swinnterton-Dyer Conjecture   128

28 Appendix C: An Actual Explanation of Elliptic Curve Addition   132

THE ERDŐS CONJECTURE ON ARITHMETIC PROGRESSIONS   143

29 Introducing Paul Erdős   144

30 Flip It: Reciprocals and Sums   147

31 Arithmetic Progressions: The Godwin's Law of Mathematics   152

32 Back To The Dynamite   157

PROBLEMS THAT ARE EASY TO STATE AND ABSOLUTELY IMPOSSIBLE TO SOLVE   161

33 Easy to Understand, Impossible to Solve   162

CONTENTS v

34 Collatz Conjecture: 1930's Version of Angry Birds   **165**

35 Goldbach Conjecture: Everything Breaks into Primes, but in a Weird Way   **171**

36 The Twin Prime Conjecture and Generalizations: Primes Parading in Pairs   **178**

37 "Perfect" Numbers: Numbers That are Way, Way Too Full of Themselves   **184**

38 Epilogue   **193**

39 Acknowledgements   **195**

40 Photo credits   **197**

Chapter 1

Introduction: Welcome to My Book!

When I tell people that I'm a research mathematician, the most common response I get is, "Wow, I hate math!" This always seemed a weird response to me, largely because I must have missed the point when it became not rude to tell other people how much you hate their vocation[1], but it's an extremely common refrain. Sometimes, this is followed by a story about how the person was good at math until some horrible occurrence, usually Algebra 2 or fractions or calculus, that tragically made them no longer like math. Apparently, part of my job as mathematician is to be a math therapist. Go figure.

The next most common response, though, is when the person tries to imagine what a research mathematician might do. This is always incredibly amusing to me because the person is almost never close. Generally, people seem to think math is either a.) something that was all figured out in the 19th century, or b.) some sort of hierarchical discipline, like karate, where the goal is to master higher and higher levels of calculus, presumably so that we can eventually challenge our masters to math combat or something. Starting from this basis to try to figure out what I do for a living means that the person either thinks I'm the next Isaac Newton or some weird math version of Jackie Chan. Unsurprisingly, neither of these is true.

For what it's worth, I think this perception of mathematics is strongly tied to how we as a country have chosen to teach math. In schools, math is taught

---

[1] "Oh, you're an architect? Wow, I hate architecture!"

in a linear fashion, where algebra 1 is followed by algebra 2 and calculus 2 begets calculus 3, and all of the material that is taught is stuff that was figured out hundreds of years ago. Nothing is ever presented as unsolved, unknown, or unanswerable, and, as a result, people think that nothing in math *is* unsolved or unanswerable. Also, I bear a striking resemblance to Jackie Chan[2].

The whole point of my rambling here, though, (as far as I remember, anyway) is that research mathematicians *do* exist, and, furthermore, we exist by the tens of thousands. We may be hard to find because we're not necessarily the most social people in the world, but we're definitely out there[3]. Believe it or not, math is a vibrant science - it has actual research, complete with all of the questions, disagreements, trends, and discoveries that pervade all of the sciences. Even more shockingly, many of the questions that we work on are surprisingly understandable (and by "understandable," I mean understandable to actual people, not mathematicians or robots or pod-people[4]).

I decided to write this book, then, to answer the question, "What, exactly, does a mathematician do?" I wanted to present some of the most important problems in mathematics today, and I wanted to do it in a way where people who wouldn't ordinarily be aware of these problems would get a chance to see what it is that gets us mathematicians so excited. I mean, let's be honest here - most mathematicians don't pick their job because they really liked addition or became enamored with improper fractions; they become mathematicians because they were introduced to problems like the ones in this book and found themselves absolutely hooked.

I also wrote this book because wanted the opportunity to make up biographical "facts" about famous mathematicians and see if they later appear on Wikipedia. I'm also hoping that I can add some quotes to the lexicon of mathematical quotations by putting them in this book and vaguely attributing them to "a famous mathematician." Of course, I'll have to depend upon you, the reader, to help me out with these. I can't vandalize Wikipedia all by myself!

---

[2]Yeah, this is false. I'm as Irish and German as a leprechaun in lederhosen.

[3]This statement actually applies for several definitions of "out there."

[4]I bring this up because I've read many a book where a mathematician had a rather warped view of what a normal person would understand.

CHAPTER 1. WELCOME TO MY BOOK! 3

1.1 Types of Mathematics: There's More than One Kind?

For what it's worth, the math world actually breaks into two main categories:

- Pure mathematicians do math problems that they think are fun, which is to say, they do math without thinking about what the applications will be. This is awesome because, since we're not constrained by trivialities like "reality," we get to play with mind-bending questions like "What if space had eleven dimensions[5]?" and "What if a times b didn't equal b times a?"

- Applied mathematicians study how math applies to real-world settings, answering questions about computers and computation, how to maximize or minimize things (profits, engine performance, etc.), how to model populations and disease spreads, and whatever other problems in life are amenable to mathematical solutions. They take the ideas that pure mathematicians come up with and find applications for them.

Pure mathematicians generally look down on applied mathematicians as lesser mathematicians for having sold out to The Man and doing easier math. Applied mathematicians, meanwhile, look down on pure mathematicians for doing their work in the land of make-believe. This petty rivalry has raged between the two branches of mathematics for hundreds of years, and it remains just as pointless and stupid today as it has ever been.

These two main categories break into any number of subcategories, most of which have names that sound frightening to the outsider (such as numerical analysis, combinatorics, linear algebra, measure theory, abstract algebra, graph theory, partial differential equations, etc.), and, quite often, to the insider as well if it's not his or her particular area. Each of these areas has its own problems and verbiage and aims and would undoubtedly make for a fine book if one felt the need to write about them.

In this book, however, I decided to focus on an area of pure math known as *number theory*, which is widely considered to be the oldest branch of mathematics. I chose number theory for four reasons:

1.) It's one of the "hot" areas of mathematics right now. Math, like

---

[5]Actually, this question has slowly crept over to the applied side of mathematics, since many physicists now study models of the universe that involve ten or eleven dimensions.

everything else on the planet, goes through cycles where some areas become more popular and others less popular, and, right now, number theory is all the rage.

2.) As noted above, it's the oldest and most storied branch of mathematics. Number theory is over 2,000 years old, and its history is replete with interesting stories, especially the ones I fabricated for this book.

3.) It has a number of really nice, accessible problems that are very good at getting people sucked in (see the chapter entitled "Easy Problems").

4.) It's my area of expertise, so I can write a book on it without having to actually do research or get off my couch.

Incidentally, if I were to rank these four reasons in order of importance, it would go 4 and then the other ones.

1.2 What Is Number Theory?

Before I begin, I want to start by saying what number theory isn't. Number theory is *not* questions like "Do numbers really exist?" or "What does it mean to be a number?" These types of questions fall more within the purview of philosophers. Or stoners. We mathematicians don't deal with that stuff around here - we're generally content to assume that numbers exist and counting works the way we think it does[6].

So, what is number theory? Number theory deals with questions about whole numbers or prime numbers. Specifically, we tend to ask questions like, "Does this equation have whole number solutions?" or "What sorts of patterns can we find about prime numbers?" Number theorists stick to nice, round, whole numbers like 3 or 17; we tend to ignore ugly numbers like π and e and $\sqrt{2}$ that do weird and unpredictable things. And don't get me started on the imaginary number i. Unfortunately, these uglier numbers often come up as we're pursuing answers to our simple questions because that's how math works, but the goal is almost always to say something about whole

---

[6]The closest thing to this that we have in math is an area known as *set theory*, where we actually give explicit definitions for things like zero and one, addition and subtraction, and all of the other standard terms and operations in math. We mathematicians decided to get these ideas squared away in the early twentieth century to make sure we didn't have to consider weird variants of math where people used different versions of, say, "zero" or "two" or "multiplication."

CHAPTER 1. WELCOME TO MY BOOK!

numbers or primes.

Why Whole Numbers?

"So, why are we so interested in whole numbers that we'd devote an entire branch of math to them?", you might be asking. Well, the origins of these sorts of questions are obvious; in the ancient world, one wouldn't talk about 2.83 sheep or $\frac{5\pi}{2}$ cows. In fact, we probably wouldn't talk about that now unless something very strange or unfortunate had happened to our livestock. There are many instances in the real world where talking about decimals or fractions doesn't make sense; we want to know if there are whole number solutions because we can only have whole numbers of objects.

Did You Just Say "In The Real World?" I Thought Pure Mathematicians Didn't Worry About The Real World Implications of What They Did.

Well, yeah. But sometimes it happens that there's overlap between what we choose to study and the real world. Besides, if we didn't pretend that there were some real world benefit to what we did, we would never get funding from outside sources.

OK, Fine. What About Primes?

Indeed. "Why are prime numbers interesting?" is a question you might be asking now. Actually, if you're not familiar with primes or you've forgotten the definition since high school, you'd probably be more likely to ask, "What are prime numbers, anyway?" That's probably a better place to begin. Let's start there:

Well, What Are They?

Prime numbers are (whole) numbers that are divisible only by themselves and 1. So, for example, 7 is prime because 7 and 1 are the only whole numbers that divide 7. 8, by contrast, isn't prime because there are other numbers that divide into 8: 2 and 4 are both examples of non-8 numbers that divide 8.

I should take a moment to clear up a common misconception here. The number 1 is generally not considered prime because the phrase "itself and 1" doesn't really make sense for 1. On the other hand, 2 is prime (which is a fact most people overlook), since its only factors are 1 and 2.[7]

Anyway, apart from the fact that it's an interesting badge of honor for a number to be prime, why do we spend so much time thinking about them? Well, primes are the building blocks of the whole numbers; after all, any whole number that isn't prime can be broken down into prime factors (for example, 30 can be broken into $2 \cdot 3 \cdot 5$, all of which are prime) but can't be broken further (since primes don't break into factors other than themselves and 1). In short, any whole number is built out of prime factors, sort of like how matter is built out of cells, or how my diet is built out of Cheez-Doodles and Coke Zero[8]. If we want to prove something about whole numbers in general, one of the best ways to proceed is to look at the building blocks that construct them; the more we know about the building blocks, the more we can say about whole numbers themselves.

Are There Other Reasons We Should Care About Primes?

There's actually a more practical reason that we're really interested in primes right now: they're really useful for computers. More specifically, prime numbers are immensely helpful in the study of encryption (i.e. encoding messages to keep them private).

Why? The basic idea for sending an encrypted message is to find an operation that's really easy to do (like locking a lock) but really hard to undo unless you have the key (like unlocking a lock). Now, let's say I gave you two tasks:

1.) Multiply 1157569866683036578989624679572 and 6075380529345458860144577398704761614649 together.

2.) 70326774294031390132703685093375489138038234246758714770 5225 244302093 is the product of two prime numbers. Find them.

The first one is easy; you take the numbers and feed them into a computer,

---

[7]Incidentally, 2, by being the only prime number that is also even, is an absolute pain to deal with and is always messing up our solutions to whatever problems we're working on. A professor of mine once noted that "if a prime number ever jumps out of the bushes and shoots you in the back of the head, it's probably 2."

[8]I eagerly await an endorsement deal here.

and then you go to lunch. Even a drooling moron could do that. Well, unless he drooled into the keyboard and caused the computer to short out. In fact, you (not the moron) could probably do this computation by hand if you had about fifteen minutes to spare.

The second one is unbelievably difficult. You take that number and feed it to your computer, and your computer will slowly turn into the moron mentioned above (minus the drool, presumably, unless you have a particularly lifelike computer), because factoring takes a long, long, LONG time and occupies way too many circuits. If you chose to actually do this by hand....God help you. Trying to do this by hand would be like trying to construct a nuclear reactor in your backyard using masking tape and Lincoln Logs.

You can see the appeal here. Taking two large prime numbers and multiplying them together is really easy; taking a large number and splitting it into primes is really, really hard. This simple idea is actually the basis for most encryption systems.

Oh.

Yeah. That's the basic idea.

1.3 Got 99 Problems But I Can Only Pick 1. Well, 9.

Now, even within the field of number theory, there are thousands and thousands of questions and unsolved problems. Of course, you can imagine that not all of these questions are of equal importance; on a list of thousands of questions, the level of importance will necessarily run the gamut from "vitally important" to "silly." This book, however, winnows the world of number theory down to nine of the most important and interesting problems that we face today. While number theorists might quibble about whether these questions are THE most important questions in number theory, there is no doubt that these are all extremely important; they're all the sorts of problems where an announcement of a solution might inspire number theorists to cancel classes for the day and go out drinking in celebration. Here are some facts about these problems that may give you a sense of just how important they are:

- *Most of these problems have been around for quite a while.* Six of the nine problems are over 100 years old. Three problems are over 250 years old, and two of them are over 1,500 years old. The newest one on the list was posed in 1985, which is new by number theory standards but still not exactly a fly-by-night deal.

- *Several of these problems have significant prize money attached.* In fact, two of the problems have million dollar bounties attached to their solutions. Another has a fifty thousand-dollar prize attached, and yet another has a five thousand-dollar prize.

- *All of these problems would have major repercussions in mathematics.* If we were to prove any one of these problems, a cascade of new results would invariably follow. In fact, there have been many, many papers written that begin with something like, "Assume the Riemann Hypothesis...." or "Assume that the ABC Conjecture is true...," so we have some sense of what sorts of results might follow from solutions to these problems.

- *All of these problems are really, really, really hard.* I don't recommend them to the amateur, but if you insist on trying them.....um, good luck.

1.4 A Word About This Book

I should probably say a few things about this book.

First of all, you should definitely buy several copies.

Second, because this book is for all levels, some of the more advanced among the readership are going to say, "Could you explain a little more about this topic?" and others are going to say, "GAAAAH! Math!"

In light of that, some of the chapters have appendices for the math inclined. The idea there was that I wanted to keep the hardcore math out of chapters themselves because the stories are easier to tell if I'm not clogging up the pages with tangled webs of equations. That said, for those who want to get a better sense of what's going on, I figured I'd throw a little bit more of the specifics of why things work and where they came from. If you're not feeling up to the appendices and just want to stick with the stories and pretty pictures, no worries - no one will force you to read every word.

Chapter 2

Verbiage

There are a few words and phrases that you're going to see repeatedly throughout the book. I figure that I should probably define them somewhere; otherwise, this tome is going to be a rather difficult read. Let's do that now:

Whole Numbers: We've already talked a bit about these, but I figured I should be a bit clearer in defining them. The whole numbers are the numbers

$$\{0, 1, 2, 3, 4, 5, 6, 7....\}.$$

Note that the ancient Greeks didn't have or understand the number zero, so we're already starting from a pretty advanced place.

What if you wanted to start with whole numbers but then expand them to include negative numbers? Well, in that case, the ancient Greeks would probably try to burn you for witchcraft. Should you survive such an assassination attempt, however, you would end up with:

Integers: These are the positive and negative whole numbers:

$$\{...-4, -3, -2, -1, 0, 1, 2, 3, 4, ...\}$$

The negatives generally make things more complicated, so we usually deal with these only if we're feeling particularly masochistic[1].

---

[1] Oddly, in math, this happens often.

I should add that if we wanted to restrict ourselves to the world of the ancient Greeks, we would stick with

Positive Integers: These are

$$\{1, 2, 3, 4, ...\}$$

Mathematicians sometimes refer to these as "counting numbers," "natural numbers," "primitive numbers," or "stupid people numbers[2]."

At some point in the book, we'll also need these:

Rational Numbers: This is one you might not have heard of, so I'll probably define it a couple more times before the book is done. "Rational numbers" is a fancy term for fractions. Anything that can be written as an integer over another integer is a rational number. So $\frac{3}{2}$ or $-\frac{9}{4}$ would work. For that matter, 2 would also count because you could write it as $\frac{2}{1}$.

I have no idea who it was that got to decide that these numbers are rational and other ones like π aren't; if you ask me, all numbers are a little crazy. The name seems to have stuck, though, so here we are.

Anyway, this isn't a dictionary, so I don't feel like spending pages and pages defining mathematical terms. Let's move on....

---

[2]These are all true except the last two.

Chapter 3

Odds and Ends

I suppose there are a couple of things I should mention before we get into the math.

To start, I feel like I should say something about what we mathematicians call a "conjecture." For those who are unfamiliar with the word, "conjecture" means something like "hypothesis," "educated guess," or "thing that feels like it should be true but no one really knows." This book is concerned with the major conjectures of number theory. For a list of these, turn back a couple of pages to the table of contents.

Having said this, it's important to note that the things we list as conjectures aren't just idle speculations; there are mounds and mounds of supporting evidence for each of these propositions. In fact, the bar for proof in mathematics is higher than it is in pretty much any other discipline. This is true because we mathematicians are, in fact, better people than those in any other discipline. HA! Just kidding. Actually, there are two reasons we do this: a.) math allows us to have the bar high, since we have the ability to actually prove things rather than just accumulating enough evidence to convince ourselves that something is true[1], and b.) because we have to be.

A great example for why we have to be careful: back in the 1700's, it was conjectured that the equation

---

[1] There are many, many stereotypes about mathematicians, and most of them are completely false. However, one of the stereotypes that is absolutely true is that if you ask a mathematician a question, he will think about the question for far longer than you expect and then give you an answer that is far more specific than you anticipated. The whole process of trying to prove something requires one to be very specific with words, and this generally carries over into real life.

CHAPTER 3. ODDS AND ENDS

$$x^4 + y^4 + z^4 = w^4$$

had no whole number solutions (unless, say, x and y are zero). Mathematicians plugged countless values into this equation, never finding a solution, and gradually accruing more and more evidence that this equation didn't have any solutions. This statement was so widely expected to be true that it even had a name; it was known as Euler's Sum of Powers Conjecture. Then, in 1988, Harvard mathematician Noam Elkies and his colleague Roger Frye discovered that

$$x = 95,800; \quad y = 217,519; \quad z = 414,560; \quad w = 422,481$$

worked, and, moreover, that this was the only solution where all of the numbers were less than a million. In other words, this is a conjecture that seemed to hold for the first *tens of thousands* of values that we plugged in, and yet it is ultimately false, thereby turning our mounds of evidence to dust. Or ashes. Or whatever it is that evidence becomes after being discredited.

All right, on to the heart of the matter...

The Riemann Hypothesis

Chapter 4

Historical Background: Straight Cash, Homey and Other Mathematical Concepts

4.1 He's Making a List, and Checking it for Money

On August 8, 1900, David Hilbert, a German mathematician with a penchant for really cool-looking hats, strode to the board at the International Congress of Mathematicians in Paris and, in an historic pronouncement, unveiled a list of ten problems that he felt would be of fundamental importance to mathematics in the 20th century. The mathematicians around the room were shocked, presumably because many of them had never seen ten problems listed in a row before[1]. Critical acclaim for the pronouncement was immediate, with some critics going so far as to call it "The talk of the new century!" and "The best list of ten problems I've ever seen!"

Encouraged by his success[2], Hilbert soon released the notes from his talk with a director's commentary in which he announced thirteen extra problems, making twenty-three in total. These problems, like their predecessors, were also met with great approval from the mathematical community.

It often happens that a work, though critically acclaimed, is quickly for-

---

[1]Plus, several of them had been studying mathematical abstractions for so long that they had forgotten how to count to ten, making Hilbert's enumeration even more shocking.

[2]Although disappointed that no one had commented on how awesome his hat looked.

CHAPTER 4. HISTORY: STRAIGHT CASH, HOMEY

gotten. This was not the case with Hilbert's problems. In fact, the result was quite the opposite, as Hilbert's list turned out to be creepily predictive of the new century. Today, seventeen of the twenty-three questions have been fully or partially resolved; moreover, for many of these questions, the solutions (or partial solutions) resulted in breakthroughs that became fundamental to the development of 20th century mathematics. Just six of the questions remain completely unsettled, although one of the questions ("Construct all metrics where lines are geodesics") is considered hopelessly broad and another question ("Is this hat awesome or what?") has been deemed by many mathematicians to be "too rhetorical to pursue."

Exactly one hundred years later, the Clay Mathematics Institute decided to honor Hilbert's historic announcement by compiling a new list of of twenty-three unsolved problems that would be of importance in the 21st century. Unfortunately, coming up with a large list of interesting unsolved problems turns out to be harder than it looks, and the Institute managed to compile just seven problems (including one - the Riemann Hypothesis - that they "borrowed" from Hilbert's original list) before they were forced to resort to rhetorical questions about hats. Realizing that a list of just seven problems - one of which wasn't even new - wouldn't exactly capture the imagination of the math community, the Clay Mathematics Institute decided to add something to the list that *would* capture the imagination: million-dollar prizes for anyone who solved any of the problems. The gambit worked; mathematicians gave the Institute's list (and money) a warm reception, and the seven problems soon came to be accepted as the natural successor to Hilbert's list.

Of all of the mathematical concepts on the Clay Mathematics list, the most interesting one for a number theorist is undoubtedly the one million dollars. However, that doesn't make for a very interesting paper, so we'll talk instead about the second most interesting one: the aforementioned Riemann Hypothesis.

Much to the amazement of his colleagues, Hilbert could count all the way up to 23.

4.2 The Riemann Hypothesis: Yeah, I'm Jealous

The Riemann Hypothesis is named after the fact that it is a hypothesis, which, as we all know, is the largest of the three sides of a right triangle. Or maybe that's "hypotenuse." Whatever. The Riemann Hypothesis was posed in 1859 by Bernhard Riemann, a mathematician who was not a number theorist and wrote just one paper on number theory in his entire career. Naturally, this single paper would go on to become one of the most influential papers in number theory history, a depressing, frustrating, and angering thought for those of us who will actually work in number theory full-time for our entire lives and will still never write a number theory paper nearly that important.

In this infuriatingly important paper, entitled *Ueber die Anzahl der Primzahlen unter einer gegebenen Grösse* (which, according to a translator I found online, translates into English as *Possibility of the Size Lower Part Primes Which Comes to Give*), Riemann introduced a mysterious new function which he called ζ (the Greek letter zeta) and asked, "When does this function hit zero?" While this may seem a really simple question, the zeta function turns out to be extremely important to the study of number theory, and Riemann's query has gone on to have far-reaching implications not just in number theory but throughout all of math and physics. What a jerk.

Of course, Riemann's work was predicated upon the earlier work of Leonard Euler. As you will soon discover, this is actually one of the recurring literary themes in this book; most of the stuff that we do in number theory is based on some idea that Euler had, so his name comes up an awful lot. Apparently, Euler was a pretty productive guy.

In this case, what Euler did was discover a new function that would give rise to this "zeta" that I mentioned above. We'll outline Euler's function and the sorts of bizarre things Riemann did to it in the next chapter.

Chapter 5
Make My Funk a Z-Func(tion)

To help us understand the function that Euler discovered, let us begin with a simple question. What happens if we take the following sum:

$$\frac{1}{1} + \frac{1}{2} + \frac{1}{3} + \frac{1}{4} + \frac{1}{5} + \frac{1}{6} + ...?$$

What happens is that the sum goes off to infinity, indicating that the above is a dumb question.

Lets try again. What happens if we take another sum:

$$\frac{1}{1^2} + \frac{1}{2^2} + \frac{1}{3^2} + \frac{1}{4^2} + \frac{1}{5^2} + \frac{1}{6^2} + ...?$$

This one is actually a little more interesting:

$$\frac{1}{1^2} + \frac{1}{2^2} + \frac{1}{3^2} + \frac{1}{4^2} + \frac{1}{5^2} + \frac{1}{6^2} + ... = \frac{\pi^2}{6}.$$

The sum

$$\frac{1}{1^3} + \frac{1}{2^3} + \frac{1}{3^3} + \frac{1}{4^3} + \frac{1}{5^3} + \frac{1}{6^3} + ...$$

is a bit of a mystery that I won't get into here (though it equals about 1.202), but this one's kind of cool:

$$\frac{1}{1^4} + \frac{1}{2^4} + \frac{1}{3^4} + \frac{1}{4^4} + \frac{1}{5^4} + \frac{1}{6^4} + ... = \frac{\pi^4}{90}$$

Euler saw all of these identities[1] and was so inspired that he asked the following question:

---
[1] Actually, he was the one who discovered them.

CHAPTER 5. Z-FUNK

The Euler Question: *Get a load of this expression*:

$$\frac{1}{1^s} + \frac{1}{2^s} + \frac{1}{3^s} + \frac{1}{4^s} + \frac{1}{5^s} + \frac{1}{6^s} + \ldots$$

Are there any s's for which interesting things happen (besides 2 and 4)?

When someone asks a good question like this, the first thing that we mathematicians do is start naming everything in the problem because naming things is way easier than solving math problems. Let's give our expression a name:

Definition: *Write the above expression as $Z(s)$, i.e.*

$$Z(s) = \frac{1}{1^s} + \frac{1}{2^s} + \frac{1}{3^s} + \frac{1}{4^s} + \frac{1}{5^s} + \frac{1}{6^s} + \ldots$$

So $Z(2) = \frac{\pi^2}{6}$, $Z(4) = \frac{\pi^4}{90}$, $Z(1)$ goes off to infinity, etc.

Now, we can restate the question above, which still doesn't change anything but again gives us the illusion of doing something productive:

The More Economical Euler Question: *What happens with $Z(s)$ when s isn't 2 or 4?*

It looks like we've succeeded in asking our question in as few characters as possible. That's progress.

Unfortunately, the answer isn't going to be quite as nice as we hoped.

Partial Answer: *There are many, many values that are completely uninteresting.*

What's the problem? Well, remember how we said that $Z(1)$ went off to infinity? It turns out that "goes off to infinity" is not an interesting thing for $Z(s)$ to do[2]. Moreover, it turns out that 1 isn't the only place where s goes off to infinity. For example, if you take any positive integer and raise it

---

[2] Okay, I suppose it's a little interesting, but compared to something like $\frac{\pi^2}{6}$? Come on. It's no contest.

CHAPTER 5. Z-FUNK   20

to the zero, you get 1. So

$$Z(0) = \frac{1}{1^0} + \frac{1}{2^0} + \frac{1}{3^0} + ... = \frac{1}{1} + \frac{1}{1} + \frac{1}{1} + ... = 1 + 1 + 1 + ...$$

which undoubtedly goes off to infinity as well.

It gets even worse if s is a negative number. Remember that having a negative exponent flips the fraction over (i.e. $\frac{1}{x^{-2}} = x^2$). So for things like $s = -1$, we have

$$Z(-1) = \frac{1}{1^{-1}} + \frac{1}{2^{-1}} + \frac{1}{3^{-1}} + ... = 1^1 + 2^1 + 3^1 + ...$$

which is also getting really, really big. This is obviously going to be a problem for any negative s. Basically, we have a function that's not even going to be defined half of the time.

From these observations, we have the following theorem to describe just how annoying and useless $Z(s)$ can be:

Major Theorem: *If $s \leq 1$ then the function $Z(s)$ blows*[3].

5.1 Why Was Euler Thinking About That Function, Anyway?

It's a funny story, actually. One day, the positive integers were minding their own business, patiently waiting in line to be added up in the Z-function, hoping that the operator hadn't chosen an s that would make them all blow up, when all of a sudden.....

....a rebel gang of numbers showed up:

$$\left(\frac{1}{1-\frac{1}{2^s}}\right)\left(\frac{1}{1-\frac{1}{3^s}}\right)\left(\frac{1}{1-\frac{1}{5^s}}\right)\left(\frac{1}{1-\frac{1}{7^s}}\right)\left(\frac{1}{1-\frac{1}{11^s}}\right)\left(\frac{1}{1-\frac{1}{13^s}}\right)....$$

---

[3]Here, by "blows," we of course mean the conventional mathematical definition of "blows up to infinity." I don't know why you would have thought I meant something else.

CHAPTER 5. Z-FUNK

Good God, thought the integers, those are the prime numbers! And they've arranged themselves in a pattern to make themselves equal to $Z(s)$!

$$Z(s) = \frac{1}{1^s} + \frac{1}{2^s} + \frac{1}{3^s} + \frac{1}{4^s} + \frac{1}{5^s} + \frac{1}{6^s}....$$
$$= \left(\frac{1}{1-\frac{1}{2^s}}\right)\left(\frac{1}{1-\frac{1}{3^s}}\right)\left(\frac{1}{1-\frac{1}{5^s}}\right)\left(\frac{1}{1-\frac{1}{7^s}}\right)\left(\frac{1}{1-\frac{1}{11^s}}\right)\left(\frac{1}{1-\frac{1}{13^s}}\right)....$$

The positive integers were aghast at the insolence of those primes. The primes had found a vehicle with which they could turn statements about integers into statements about prime numbers. And they had done it by hijacking the integers' beloved $Z(s)$, too!

While the non-prime integers may still harbor some resentment over this coup, we number theorists view this revolution in much the same way Americans view the American Revolution. If we want to think about questions related to prime numbers, we can use this $Z(s)$ to translate them to questions about regular old positive integers, which are much easier to deal with[4]. Since number theorists are obsessed with primes, this is, as Euler so eloquently noted, a "big freaking deal."

---

[4]Don't believe me that positive integers are easier than primes to deal with? Okay, smarty pants, answer me these questions: What's the next prime number after 7549? What's the next positive integer after 7549? Which question was easier? Yeah, that's what I thought.

Chapter 6

The Zeta Function: Magical, Mystical, and....Dear God, What Is That Thing?

Despite the fundamental importance of $Z(s)$, Euler's efforts to tame this function and domesticate it and maybe make it do tricks were hamstrung by the fact that it always seemed to be blowing up at inopportune times. As a result, he gave up and spent the remainder of his life going blind working on other mathematics in a dark attic.

The function continued to lay prostrate in its useless state for over half of a century until Bernhard Riemann came along. He took the function into his office, watched it self-destructively blow up any time a negative number was mentioned, and decided that it needed help. Although Riemann was not a number theorist by trade, he felt, much like that guy in the movie "Lorenzo's Oil," that he could teach himself enough number theory to cure Z of its horrible ailment.

CHAPTER 6. ZETA: THE MAGIC AND THE HORROR

Bertrand Riemann: Hypothesizin'

After months in his basement laboratory[1], Riemann emerged with what he thought was a remedy: a new function that he called ζ (the Greek letter zeta) because Riemann mistakenly thought that he was Greek. It was a function that had the scientific importance of the Frankenstein monster[2] and the aesthetic appeal of, well, the Frankenstein monster. Take a look at this thing:

$$\zeta(s) = \frac{1}{1-2^{1-s}} \sum_{n=0}^{\infty} \frac{1}{2^{n+1}} \sum_{k=0}^{n} (-1)^k \frac{n!}{k!(n-k)!} (k+1)^{-s}.$$

It's hideous[3]!

---

[1]Riemann probably didn't have a basement laboratory. He probably had a cushy university office with a comfy couch where he would lie down and sometimes take naps, and then sometimes people would knock on his office door and he would groggily tell them that he wasn't napping but instead "thinking about mathematics." Believe me, that little trick isn't fooling anybody, Riemann.

[2]I suppose that the Frankenstein monster didn't have any actual scientific importance since it was a fictional monster, but the scientists in the movie all looked pretty impressed, so that's close enough for me.

[3]Sometimes, in an attempt to get the same amount of shock value in less space, mathematicians will consolidate notation and write the function as

$$\zeta(s) = \frac{\Gamma(1-s)}{2\pi} \oint_\gamma \frac{u^{z-1}}{e^{-u}-1} du.$$

CHAPTER 6. ZETA: THE MAGIC AND THE HORROR   24

Why is this a fix, you ask? Well, let's take it for a test drive and try out some values. First, let's try $s = 2$:

$$\zeta(2) = \frac{\pi^2}{6}.$$

Wait, we've seen that value before. That's $Z(2)$!

Now, let's do another one:

$$\zeta(4) = \frac{\pi^4}{90}.$$

That's $Z(4)$!

We can try value after value for s, but the same result will keep happening:

Riemann's Result: *For any s for which $Z(s)$ doesn't blow up, $\zeta(s) = Z(s)$.*

Riemann had found a function that mirrored $Z(s)$. Unlike $Z(s)$, though, $\zeta(s)$ didn't blow up if s was less than 1. And ζ was prepared to handle all kinds of numbers! Fractions! Decimals! Imaginary numbers like $\sqrt{-1}$! Combinations of real numbers and imaginary numbers! $\zeta(s)$ was like a post-spinach-Popeye version of $Z(s)$.

Unfortunately, like Achilles, ζ still had one flaw. There was one single value for s that ζ couldn't handle:

Fundamental Statement About Zeta: $\zeta(1)$ *is undefined.*

Oh well. You can't have everything.

6.1 Getting Back To Our Original Question

Now that we've found a suitable replacement for the mercurial $Z(s)$, we can try asking Euler's question again:

The Euler Question (again): *Is there anything interesting about $\zeta(s)$?*

For those who are interested, there's a more thorough explanation of how Riemann got from $Z(s)$ to $\zeta(s)$ in Appendix A.

CHAPTER 6. ZETA: THE MAGIC AND THE HORROR

This question can often be a dog-whistle type question for mathematicians; when we say, "Is it interesting?", we often mean, "Does it hit zero a lot?" That may not be interesting to everyone, but mathematicians think zero is really, really fascinating. We're kind of like that kid with the video camera in "American Beauty" who thinks that a bag blowing in the wind is the most beautiful thing in the world; you may think we're weird, but, well, who asked you anyway?

Since "interesting" means different things to different people anyway, let's try thinking about this zero stuff and see if it gets us anywhere:

The Euler Question (yet again): *When does $\zeta(s)$ hit zero?*

There are actually quite a few places where it is easy to show it hits zero:

Partial Answer: $\zeta(s) = 0$ *when* $s = -2, -4, -6, -8, -10, ...$

OK. We're partway there. Those are the ones that are actually pretty easy to find. What about other ones?

The Euler Question (for the last time, I swear): *Okay, okay. Besides negative even integers, when does $\zeta(s)$ hit zero?*

We don't have all the answers to this question, but here's one[4]:

$$\zeta(\frac{1}{2} + 14.134725142i) = 0.$$

Here's another one:

$$\zeta(\frac{1}{2} + 21.022039639i) = 0.$$

And another:

$$\zeta(\frac{1}{2} + 25.010857580i) = 0.$$

And yet another:

$$\zeta(\frac{1}{2} + 30.424876126i) = 0.$$

Starting to notice a pattern? They all seem to be $\frac{1}{2}$ plus some multiple of i.

---
[4]In case you've forgotten, $i = \sqrt{-1}$.

It would be natural to ask whether this pattern will continue. It's so natural to ask, in fact, that Riemann beat you to it by over 150 years.

The Riemann Question: *Let's ignore those negative even integers for now. If $\zeta(s) = 0$, does that mean that s is $\frac{1}{2}$ plus a multiple of i?*

This question has a less than satisfactory answer:

The Riemann Answer: *I have no earthly idea.*

Coming up with a more satisfactory answer (such as "yes" or possibly even "no") is such a difficult thing to do that a $1,000,000 reward has been promised to the person who finally does. The best we currently have is the following:

The Riemann Guess: *The answer looks like it should be yes (?)*

or, in mathier speak:

The Riemann Hypothesis (The Riemann Guess with Fancier Words): *If $\zeta(s) = 0$ and s is not a negative even integer then s is $\frac{1}{2}$ plus a multiple of i.*

or, in even mathier speak:

The Riemann Hypothesis (Take 2): *If $\zeta(s) = 0$ and s is not a negative even integer then $s = \frac{1}{2} + it$ for some real number t.*

The Riemann Hypothesis is considered by many mathematicians to be the most important unsolved problem in mathematics today.

6.2 Wait, Wait, That's It? The Question of When Some Esoteric Function Hits Zero is The Most Important Problem In Math?

Yep.

6.3 How?

I'm glad you asked. Because the Riemann Zeta Function is based on such a simple equation (namely, $Z(s)$), it's something that comes up in a lot of computations, so having a good understanding of it would help us calculate all sorts of interesting properties about integers and prime numbers. As far as conjectures go, it's not as sexy (or as likely to generate crank mail) as something like the Twin Primes Conjecture, but we all know that beauty is skin-deep, and the Riemann Hypothesis is nothing if not deep[5].

6.4 Can You Give An Example of Something That the Riemann Hypothesis Can Show Us?

Funny you should ask. The next chapter is all about one of the most famous examples: Gauss' Prime Number Theorem.

---

[5]Heck, it took me like ten pages to define the stupid thing. That's pretty deep.

Chapter 7

Application: Primes on Parade

7.1 Boredom is the Mother of Invention

Let us begin with a question which is actually three questions:

Question(s): *How many prime numbers are there up to 100? How about up to 1,000? Or 1,000,000?*

One way of answering the above is to manually count all of the primes up to 100 or 1,000 or 1,000,000. I would do that, except that it sounds like a lot of work and would be kind of boring, and, besides, the Red Sox game is on. So that's out.

We are forced, then, to rephrase our question:

Better Question: *Counting is boring, and I'm lazy. Is there some formula that I could use where I just plug in some number and it will do all the work for me?*

This was the question that a fifteen-year old named Carl Friedrich Gauss considered in 1792. Unlike me, Gauss did not have an urgent Red Sox game to attend to, so he sat down, looked through the data, and came up with a pretty good answer:

Gauss' Answer: *The number of primes less than x is about $\frac{x}{\ln x}$.*

CHAPTER 7. APPLICATION: PRIMES ON PARADE

Gauss was not a baseball fan.

So if we wanted to know the number of primes less than 1,000, we could just calculate $\frac{1,000}{\ln 1,000}$ and we've got a pretty reasonable estimate.

This work should have been impressive enough for somebody who wasn't yet old enough to drive[1]. However, Gauss wasn't satisfied and announced, "No! I can do even better! I shall come up with a function that comes even closer to the correct number of primes! And the function will be easy to calculate!"

Although this sounded like a bunch of empty promises to the wary public in an election year like 1792, Gauss actually succeeded in his quest and found exactly the function he was looking for. He called this function Li because he, like most 18th century number theorists, was a big fan of kung-fu legend Bruce Lee[2].[3] Since Li actually turns out to be pretty straightforward to calculate, if we want to find the number of primes up to 1,000, we can just

---

[1] Although I suppose this was less of an issue before the invention of the automobile.

[2] Oddly enough, Li is also the first letters of the words "Logarithmic Integral." Coincidence? Probably. I'm still going with the Bruce Lee explanation for the name.

[3] If you're curious, the actual definition of Li is given by

$$Li(x) = \int_2^x \frac{dt}{\ln t}.$$

If you weren't curious, well, too bad. It's your fault for looking down here at the footnote in the first place.

calculate $Li(1,000)$ and we've got an even better estimate than before[4]. This realization deserves bold letters:

The Prime Number Theorem: *The number of primes less than x is approximately $Li(x)$.*

7.2 Riemann Hypothesis: Karate Kicking Li's Error Terms Since 1859

In the last section, we made a lot of nebulous statements like, "This is a good estimate," or "This is an even better estimate." This is unfortunate, as mathematicians don't like ambiguously descriptive words like "good" because they don't really tell us anything. We are forced, then, to ask the question, "How good of an estimate is this, anyway?" Annoyingly, this question also uses the word "good," but we'll let it slide because it's a useful question; in fact, it is such an important question that it deserves italics:

The Li Question: *How closely does Li estimate the number of primes less than x?*

As before, we start by naming things. The most obvious candidate for a makeover is "the number of primes less than x," which is annoying to write out all the time. To help with this, I'm going to call this quantity $P(x)$.[5] Armed with this notation, we rephrase the question and pretend that we've accomplished something in doing so:

The More Economical Li Question: *How far apart do $Li(x)$ and $P(x)$ get?*

---

[4]It should be noted that while Gauss came up with these guesses, they weren't proven to be correct until 1896. In other words, instead of taking the five minutes to count the number of primes up to 1,000, mathematicians spent hours coming up with a guess for what the answer should be, then spent over a hundred years proving that the guess was correct. Yep, that was productive.

[5]Mathematicians actually call this $\pi(x)$; however, since we already have $\pi = 3.14159....$, I didn't want to confuse the readership.

CHAPTER 7. APPLICATION: PRIMES ON PARADE

A reasonable approximation for Li(x)

Sometimes, we call the difference between $Li(x)$ and $P(x)$ the *error term in the Prime Number Theorem*. Sometimes, we don't. Really, it just depends how we're feeling.

In our pre-Riemann Hypothesis world, the answer to the above question was unsatisfying:

The Li "Answer": *Hopefully, not very.*

In fact, we know that $Li(x)$ and $P(x)$ can differ by as much as $\sqrt{x} \cdot ln\ x$ because we've seen it in actual data. The hope is that they don't differ by much more than that because $\sqrt{x} \cdot ln\ x$ is pretty small relative to the number of things we're counting, which would mean that the Li function does a really good job of approximating $P(x)$. In the last hundred years, mathematicians have come up with better and better answers, but we're still nowhere near where the data indicates that we should be.

On the other hand, if the Riemann Hypothesis were true, we would have a very, very good answer to this question:

The Li Answer with Riemann: *If the Riemann Hypothesis is true then $Li(x)$ and $P(x)$ never differ by more than about $\sqrt{x} \cdot \ln x$.*

In other words, what we hope to be true is actually true if the Riemann Hypothesis is correct. That's a pretty powerful hypothesis.

Our story would normally end here, except that there's actually a weird sidenote to this:

The *Li* Answer with Riemann, but Backwards: *If $Li(x)$ and $P(x)$ never differ by more than about $\sqrt{x} \cdot \ln x$ then the Riemann Hypothesis is true.*

So as it turns out, the question about the error term and the Riemann Hypothesis are *actually the same question*; if you prove one of them, the other is necessarily true.

Man, math is weird sometimes.

Chapter 8

Appendix A: Analytic Continuation

In this section, I'll discuss a little more of the math that goes on. If you aren't really interested in the math behind it, then.....wait, seriously? You spent the last twenty pages reading about the Riemann Hypothesis but don't care about math? What kind of garbage is that? Suck it up and keep reading.

8.1 Riemann's Number Theoretical Patch: Better than Number Theoretical Gum

Let's go back to Riemann's idea. When we last left our hero, he had stumbled upon the following creature:

$$Z(s) = \frac{1}{1^s} + \frac{1}{2^s} + \frac{1}{3^s} + \frac{1}{4^s} + ...$$

This function was having a hard time getting through the world because it exploded at the mere mention of numbers less than or equal to 1. Riemann felt that this function would benefit from some sort of patch, so he decided to try multiplying it by $(1 - \frac{2}{2^s})$.[1] Armed with this patch, Z became a little less temperamental:

$$(1 - \frac{2}{2^s})Z(s) = \frac{1}{1^s} - \frac{1}{2^s} + \frac{1}{3^s} - \frac{1}{4^s} + \frac{1}{5^s} - \frac{1}{6^s}....$$

---

[1]This is quite similar to the movie "Patch Adams," wherein Patch fixes his friend's leaky cup by multiplying it by $(1 - \frac{2}{2^s})$.

CHAPTER 8. APPENDIX A: ANALYTIC CONTINUATION 34

The right hand side of this function is now defined when $s > 0$. That's progress.

How does this help? Well, Riemann decided to divide both sides of the above by $(1 - \frac{2}{2^s})$. This gives

$$Z(s) = \frac{\frac{1}{1^s} - \frac{1}{2^s} + \frac{1}{3^s} - \frac{1}{4^s} + \frac{1}{5^s} - \frac{1}{6^s}\cdots}{1 - \frac{2}{2^s}}.$$

Now, he had an expression for Z that was defined for s's all the way down to zero. He decided to celebrate in traditional mathematics fashion by renaming his function; from now on, this Z would become known as......ζ:[2]

$$\zeta(s) = \frac{\frac{1}{1^s} - \frac{1}{2^s} + \frac{1}{3^s} - \frac{1}{4^s} + \frac{1}{5^s} - \frac{1}{6^s}\cdots}{1 - \frac{2}{2^s}}.$$

8.1.1 An Obvious Symmetry

All right, so we've gotten s down to 0 instead of 1. How do we get s to go the rest of the way?

Well, as it turns out, if s is between 0 and 1, $\zeta(s)$ has a very nice symmetry to it. In fact, I'm sure you noticed this symmetry immediately, so I don't need to point it out, but I'll do so anyway:

$$\zeta(s) = 2^s \pi^{s-1} \sin\left(\frac{s\pi}{2}\right) \left[\int_0^\infty e^{-y} y^{-s} dy\right] \zeta(1-s)$$

when s is between 0 and 1. That was obvious.

Now, note that if you plugged in, say, $\frac{1}{4}$ for s, you end up with $\zeta(\frac{3}{4})$ on the right-hand side. This isn't much of an improvement, because we can already calculate $\zeta(\frac{1}{4})$, so we basically just took something we knew and made it harder.

But Riemann had another thought. What if you plugged in, say, -3 for s? Then the above becomes:

$$\zeta(-3) = 2^{-3} \pi^{-4} \sin\left(\frac{-3\pi}{2}\right) \left[\int_0^\infty e^{-y} y^3 dy\right] \zeta(4).$$

---

[2]Note that ζ is still not defined for $s = 1$ because if we plugged in 1 for s, the bottom would be zero and division by zero is bad. Very, very bad.

CHAPTER 8. APPENDIX A: ANALYTIC CONTINUATION

This is ugly. However, it's doable. Or rather, the right side is doable; $\zeta(4)$ can be easily calculated, and there's no reason you couldn't calculate the rest of the things on the right hand side as well[3]. But wait - we've got an equal sign, and the thing on the right is defined - the thing on the left is defined, too (and equals the thing on the right). So $\zeta(-3)$ is finally defined! And this definition doesn't even involve the word "blow."

What's special about -3? Nothing! We could have done this for any negative number. In other words, if I wanted to find $\zeta(s)$ for some negative value of s, I could simply plug this negative value into the expression above; the right-hand side would give me a bunch of stuff I could calculate, including ζ of some positive number (which has already been defined). Armed with this trick, we can now evaluate ζ for anything - except, of course, $s = 1$.[4]

And thus, the Z-function was made whole.

---

[3] Note that I say that *you* can calculate it because I certainly have no interest in doing so.

[4] In case you're wondering, mathematicians have a word for a function like this that works nicely at every possible s except for some limited number of values (in this case, one); that word is *meromorphic*. They also have some words for the points where the function isn't defined, but those words are a bit too colorful for this book and need not be repeated here.

The Generalized Riemann Hypothesis

Chapter 9

Generalized Riemann Hypothesis: Because The Riemann Hypothesis Wasn't Hard Enough

9.1 How Much Harder Can We Make this Stupid Thing, Anyway?

In math, we have a very important saying, which goes, "If at first you don't succeed, find ways to make the problem even harder so that you feel even more hopeless." This quote was probably famously said by Euclid or possibly Euler. Or perhaps it was Einstein. Or Eminem. I always get those E's confused.

Nowhere has this credo had more success than in the topic of the current chapter, wherein mathematicians took the Riemann Hypothesis, a problem that is so hard that it has appeared on not one but *two* turn-of-the-century lists of the hardest and most important problems in mathematics, and decided to make it more general so that it would be even harder to solve. Don't ask me why they did this; it was apparently seen as a good idea at the time.

So how do you make the Riemann Hypothesis even harder? Well, let's go back to how we defined it in the first place.

CHAPTER 9. GENERALIZED RIEMANN

9.2 A Brilliant Insight and a Long Name

Remember Euler's favorite function, the Z-function? Of course you do. It was only about ten pages ago. It looked like this:

$$Z(s) = \frac{1}{1^s} + \frac{1}{2^s} + \frac{1}{3^s} + \frac{1}{4^s} + \frac{1}{5^s} + \frac{1}{6^s} + \ldots$$

Well, a 19th century Belgian mathematician named Johann Peter Gustav Lejeune Dirichlet[1] came along and said, "The tops of all of these fractions are boring. They're all 1! Sure, the bottoms are interesting, but the tops are a barren wasteland. What if we put in some sort of a function on top? Also, my name is too long."

Ever the aesthete, Dirichlet took it upon himself to make the tops of the fractions (or "numerators", for those of you who remember your middle school arithmetic) more appealing to the casual mathematician. Of course, Dirichlet didn't want to make them *too* interesting, as he realized that something like

$$Z(s) = \frac{H^1(Gal(\bar{k}/k), Pic(X_{\bar{k}}))}{1^s} + \frac{\chi(z \int_{\mathbb{Z}_p} \psi_\gamma(f(\frac{x\eta}{7\pi^2}))dx)}{2^s}$$
$$+ \frac{Proj\left(\bigoplus_{d\geq 0} H_0(X^0, d(K_{\tilde{X}} + \sum_{k=1}^{dim(X)} \sqrt[k]{2}\frac{k}{c(k)}))\right)}{3^s} + \ldots$$

would give mathematicians nightmares for centuries to come. Instead, he decided to focus on numerators with simple patterns that would be recognizable to mathematicians of all stripes. He started with the most basic pattern he could think of:

$$Z(s) = \frac{1}{1^s} + \frac{0}{2^s} + \frac{1}{3^s} + \frac{0}{4^s} + \frac{1}{5^s} + \frac{0}{6^s} + \frac{1}{7^s} + \frac{0}{8^s} + \ldots$$

"That's better," thought Dirichlet. Then, he started to get more creative, realizing that instead of just having a pattern that alternates between two numbers, he could have a pattern that repeated every three terms:

$$Z(s) = \frac{1}{1^s} + \frac{-1}{2^s} + \frac{0}{3^s} + \frac{1}{4^s} + \frac{-1}{5^s} + \frac{0}{6^s} + \frac{1}{7^s} + \frac{-1}{8^s} + \frac{0}{9^s} + \frac{1}{10^s} + \frac{-1}{11^s} + \frac{0}{12^s} + \ldots$$

---

[1] I'm not kidding. That was his name. Signing official documents used to take him weeks.

CHAPTER 9. GENERALIZED RIEMANN

or four terms:

$$Z(s) = \frac{1}{1^s} + \frac{0}{2^s} + \frac{-1}{3^s} + \frac{0}{4^s} + \frac{1}{5^s} + \frac{0}{6^s} + \frac{-1}{7^s} + \frac{0}{8^s} + + \frac{1}{9^s} + \frac{0}{10^s} + \frac{-1}{11^s} + \frac{0}{12^s} + \ldots$$

or even five terms:

$$Z(s) = \frac{1}{1^s} + \frac{i}{2^s} + \frac{-i}{3^s} + \frac{-1}{4^s} + \frac{0}{5^s} + \frac{1}{6^s} + \frac{i}{7^s} + \frac{-i}{8^s} + \frac{-1}{9^s} + \frac{0}{10^s} \ldots$$

Dirichlet was pleased with his artistry, and he decided (as many artists do) to display his handiwork in simple, yet elegant, picture frames around his home and office. He was in the process of hanging one of these frames on his bedroom wall when he was struck with a stunning realization:

"Wait a minute," exclaimed Dirichlet. "If the original Z-function was able to tell interesting things about the primes, perhaps these new variants with the patterns on top should be able to tell us interesting things about patterns within the primes!"

He also began to realize that most people would consider it weird to hang equations on the wall and that from a social perspective, he might want to consider refocusing his artistic efforts on things like, say, flowers. But that's a story for another chapter. In this chapter, we'll focus mostly on $Z(s)$.

9.3 Producing Properties of Primes

The driving force behind Dirichlet's mathematical realization came from the fact (which Dirichlet discovered) that his new functions could be written entirely in terms of products of primes.

I'll explain what I mean. Think back to the the last chapter, and you'll recall that there was a nice, prime-y way of writing Euler's function $Z(s)$:

$$Z(s) = \frac{1}{1^s} + \frac{1}{2^s} + \frac{1}{3^s} + \frac{1}{4^s} + \frac{1}{5^s} + \frac{1}{6^s} \ldots$$
$$= \left(\frac{1}{1-\frac{1}{2^s}}\right)\left(\frac{1}{1-\frac{1}{3^s}}\right)\left(\frac{1}{1-\frac{1}{5^s}}\right)\left(\frac{1}{1-\frac{1}{7^s}}\right)\left(\frac{1}{1-\frac{1}{11^s}}\right)\left(\frac{1}{1-\frac{1}{13^s}}\right)\ldots$$

Well, Dirichlet noticed that the same sort of decomposition into primes could

CHAPTER 9. GENERALIZED RIEMANN

be done with each of the alterations of $Z(s)$ that he had written down. For instance, if we take the pattern that repeated every three terms, we have

$$Z(s) = \frac{1}{1^s} + \frac{-1}{2^s} + \frac{0}{3^s} + \frac{1}{4^s} + \frac{-1}{5^s} + \frac{0}{6^s} + \frac{1}{7^s} + \frac{-1}{8^s} + \frac{0}{9^s} + \frac{1}{10^s} + \frac{-1}{11^s} + \frac{0}{12^s} + \ldots$$

$$= \left(\frac{1}{1 - \frac{(-1)}{2^s}}\right)\left(\frac{1}{1 - \frac{0}{3^s}}\right)\left(\frac{1}{1 - \frac{(-1)}{5^s}}\right)\left(\frac{1}{1 - \frac{1}{7^s}}\right)\left(\frac{1}{1 - \frac{(-1)}{11^s}}\right) \ldots$$

Now, you might be wondering why, on the second line, some of the primes (2^s, 5^s, etc.) have a -1 above them and some (like 7^s) have a 1. The answer is, "Because the first line told us that we should," and, of course, we must always listen when numbers speak. In other words, since the first line had $\frac{-1}{2^s}$, the second line will also have $\frac{-1}{2^s}$; similarly, since the first line had a $\frac{1}{7^s}$, the second line does as well. (Note that this also holds true for $\frac{0}{3^s}$, which appears on the first and second lines.)

A similar decomposition could be done with Dirichlet's other patterns as well; for example:

$$Z(s) = \frac{1}{1^s} + \frac{0}{2^s} + \frac{-1}{3^s} + \frac{0}{4^s} + \frac{1}{5^s} + \frac{0}{6^s} + \frac{-1}{7^s} + \frac{0}{8^s} + \frac{1}{9^s} + \frac{0}{10^s} + \frac{-1}{11^s} + \frac{0}{12^s} + \ldots$$

$$= \left(\frac{1}{1 - \frac{0}{2^s}}\right)\left(\frac{1}{1 - \frac{-1}{3^s}}\right)\left(\frac{1}{1 - \frac{1}{5^s}}\right)\left(\frac{1}{1 - \frac{-1}{7^s}}\right)\left(\frac{1}{1 - \frac{-1}{11^s}}\right)\left(\frac{1}{1 - \frac{1}{13^s}}\right) \ldots$$

and

$$Z(s) = \frac{1}{1^s} + \frac{i}{2^s} + \frac{-i}{3^s} + \frac{-1}{4^s} + \frac{0}{5^s} + \frac{1}{6^s} + \frac{i}{7^s} + \frac{-i}{8^s} + \frac{-1}{9^s} + \frac{0}{10^s} \ldots$$

$$= \left(\frac{1}{1 - \frac{i}{2^s}}\right)\left(\frac{1}{1 - \frac{(-i)}{3^s}}\right)\left(\frac{1}{1 - \frac{0}{5^s}}\right)\left(\frac{1}{1 - \frac{i}{7^s}}\right)\left(\frac{1}{1 - \frac{1}{11^s}}\right)\left(\frac{1}{1 - \frac{-i}{13^s}}\right) \ldots,$$

and of course,

$$Z(s) = \frac{1}{1^s} + \frac{0}{2^s} + \frac{1}{3^s} + \frac{0}{4^s} + \frac{1}{5^s} + \frac{0}{6^s} + \frac{1}{7^s} + \frac{0}{8^s} + \frac{1}{9^s} + \frac{0}{10^s} + \frac{1}{11^s} + \frac{0}{12^s} + \ldots$$

$$= \left(\frac{1}{1 - \frac{0}{2^s}}\right)\left(\frac{1}{1 - \frac{1}{3^s}}\right)\left(\frac{1}{1 - \frac{1}{5^s}}\right)\left(\frac{1}{1 - \frac{1}{7^s}}\right)\left(\frac{1}{1 - \frac{1}{11^s}}\right)\left(\frac{1}{1 - \frac{1}{13^s}}\right) \ldots$$

In the Riemann Hypothesis chapter, we saw that this decomposition allowed $Z(s)$ to be treated like a Rosetta Stone that could translate questions about primes into questions about regular old integers. Now that we can do the same sort of decomposition for these variants of $Z(s)$, we have the math equivalent of a a small arsenal of Rosetta Stones that we can use to translate other questions about primes. Of course, the phrase "arsenal of Rosetta Stones" seems kind of weird, since there was only one Rosetta Stone, and the other stones probably would have been called something else, and also because archaeologists only needed one Rosetta Stone to translate ancient Egyptian hieroglyphs and hence the other stones would have been effectively redundant, but it was the best analogy I could think of. If you don't like it, write your own book.[2]

9.4 Longer Patterns?

Having discovered this fortuitous intersection between aesthetics and brilliant mathematical questions, Dirichlet decided to push further, and he asked whether one could find useful patterns that repeated every 6 terms, or every 7 terms, or possibly even every 8 terms. Or 9. Or perhaps 10. Dirichlet soon realized that the answer to all of these questions was "yes"; in fact, the work he had done to create patterns of length 2 through 5 could easily be generalized to make a much more far-reaching statement:

Dirichlet's Discovery: *There are lots and lots and lots and lots of patterns that you could put into the numerators of $Z(s)$ that would give you interesting*

---

[2]Tangentially related yet awesome aside: at the beginning of the book, I mentioned that one of the major modern applications of number theory is in cryptanalysis (i.e. codewriting and codebreaking). Well, over the last 100 years, scholars have realized that there's a huge overlap between breaking codes and trying to decipher ancient languages and scripts, since they're basically asking the same sorts of questions (how do I take this encoded thing and make it readable?). As a result, many of the mathematical methods that are used in codebreaking are now used to translate dead languages. In fact, the most important decipherment in the last century, the decoding of Linear B (an ancient version of Greek) owes its success as much to cryptanalytic methods as it does to any actual archaeology or historical research and was completed by a cryptanalyst who had served as a codebreaker for the British during World War II.

I certainly can't do the topic justice in a footnote, but if you're interested, *The Code Book* by Simon Singh does an awesome job telling the history of codebreaking and has a whole section devoted to its use in archaeology.

CHAPTER 9. GENERALIZED RIEMANN

statements about the primes.

In fact, there is at least one (and usually several) pattern(s) of any length you want.

So if I wanted to find a pattern that repeated every, say, 15 terms and says something useful about the primes, not only am I guaranteed to be able to find one (by Dirichlet's realization), I should probably be able to find several. (In fact, there are four such patterns.)

As a result of this discovery, Dirichlet now knew that he would have many, many functions that he could spend his time studying and admiring artistically. The key, though, is that *every single one* of these functions is important; each one of the functions splits up nicely into a product (just like the original $Z(s)$ did), and each one tells something new and interesting about prime numbers.

9.5 A Small Issue

Naturally, Dirichlet was pleased with his new creations, and he began to congratulate himself on his efforts until he discovered that he had two very obvious problems. First, there are many choices for the sorts of patterns you could place in the numerators, but the notation "$Z(s)$" doesn't really indicate which one you've chosen, so if I wrote, "Consider the function $Z(s)$...", you'd have no idea which pattern I was using and, hence, no idea what the heck I was talking about. Secondly, Euler's lawyers had called Dirichlet to inform him that the expression "$Z(s)$" was copyrighted by Euler's estate, and if Dirichlet chose to go ahead and call his new functions $Z(s)$ then Prussian law dictated that Euler "would be compelled to rise out of his grave and strangle Dirichlet."

Realizing the difficulty of negotiating with a dead guy and the absurdity of having a function that could actually be many different functions, Dirichlet decided that he should rename his function and add on some sort of marker to indicate which n he was using. As such, he started using L instead of Z, and he began to call his functions $L(s, \chi_n)$, which means "the L function where the pattern in the numerators repeats every n terms. And χ (the Greek letter "chi," pronounced "kai") takes care of the mathematicians' compulsive need to add a Greek letter to everything." In other words,

$$L(s,\chi_3) = \frac{1}{1^s} + \frac{-1}{2^s} + \frac{0}{3^s} + \frac{1}{4^s} + \frac{-1}{5^s} + \frac{0}{6^s} + \frac{1}{7^s} + \frac{-1}{8^s} + \frac{0}{9^s} + \frac{1}{10^s} + \frac{-1}{11^s} + \frac{0}{12^s} + \ldots$$

and

$$L(s,\chi_4) = \frac{1}{1^s} + \frac{0}{2^s} + \frac{-1}{3^s} + \frac{0}{4^s} + \frac{1}{5^s} + \frac{0}{6^s} + \frac{-1}{7^s} + \frac{0}{8^s} + \frac{1}{9^s} + \frac{0}{10^s} + \frac{-1}{11^s} + \frac{0}{12^s} + \ldots,$$

and of course

$$L(s,\chi_5) = \frac{1}{1^s} + \frac{i}{2^s} + \frac{-i}{3^s} + \frac{-1}{4^s} + \frac{0}{5^s} + \frac{1}{6^s} + \frac{i}{7^s} + \frac{-i}{8^s} + \frac{-1}{9^s} + \frac{0}{10^s} + \frac{1}{11^s} + \frac{i}{12^s} + \ldots$$

9.6 Enough With The Definitions - Do These Do Anything Cool?

Well, there are a couple of instances where the answer is yes. For instance, there's this one:

$$L(2,\chi_5) = \frac{4\sqrt{5}}{125}\pi^2.$$

Any time you can get π^2 showing up, that's pretty cool.

There's also this one:

$$L(2,\chi_{12}) = \frac{\sqrt{3}}{18}\pi^2.$$

There are several more values for which interesting things happen, too. Let's be honest, though; when we say, "Does this do anything cool?", we mean, "Does this hit zero a lot, and, if so, where?" In other words, we're looking to see if something like the Riemann Hypothesis shows up here, too.

9.7 Well, Does It?

Well, when we asked that question last time, we immediately ran into the problem where the function started blowing up.

9.8 Ah, Yes. Back To Riemann's Lab!

Right. For every choice of n, $L(s, \chi_m)$ blows up for $s \leq 0$. Some of them even blow up for $s \leq 1$. We'll have to address that first.

Chapter 10

Let's Get Ready to Rumble and Rewrite the Riemann Hypothesis!

10.1 Hurwitz and Piltz: Rumbling to the Rescue

While Johann Peter Gustav Lejeune Herbert Walker Iodine Shaquala Dumbledorf Hussein Dirichlet could well have used his new function to generalize Riemann's Hypothesis, he declined to do so, citing the fact that he was, quote, "already dead by the time that Riemann had formulated his hypothesis." Because of this act of cowardice on Dirichlet's part, it was left to another pair of mathematicians, the infamous tag team of Adolf Hurwitz and Adolf Piltz (or The Amazing Adolfs, as they were known to fans at the time), to finally generalize the Riemann Hypothesis so that these Dirichlet functions would be included, too.

Hurwitz and Piltz were an interesting pair. Hurwitz was the brainier of the two, as he was particularly facile with all different types of number theory and analysis and, as a result, has several important functions named after him today. Piltz, by contrast, was more of the brawn behind the operation, often overpowering opponents before forcing them into submission with his patented "Complex Integration" finishing move. Together, they proved to be an unstoppable force in the math world; in 1893, they would go on to win the Mathematics Tag Team Championship Belt from fellow mathematicians

CHAPTER 10. READY FOR RIEMANN! 47

David Hilbert and Herman Minkowski[1] when, in one of the most famous tag team matches in mathematics history, Hurwitz discovered a new automorphism theorem for surfaces with genus larger than one and Piltz kicked Minkowski in the face[2].

10.2 Puttin' on the Patch. Again

It was Hurwitz, the smarter of the pair, who found these $L(s, \chi_n)$'s that Dirichlet had unleashed, and, much as Riemann before him, took them into his office and tried to get them to stop blowing up. After much work and many failed attempts, Hurwitz finally discovered his function that would act as a patch; it would act like $L(s, \chi_n)$ everywhere that the $L(s, \chi_n)$ was defined, but it wouldn't be nearly as temperamental. Unfortunately, Hurwitz was so excited by his discovery of a patch that he forgot to come up with another name for it, so we now call this patched function "$L(s, \chi_n)$" (which, as you might confusedly remember, is the same name as the function that Hurwitz was trying to patch in the first place).

This new version of the L-function that Hurwitz came up with was no prettier than Riemann's ζ function, since it looks like

$$L(s, \chi_n) = \frac{1}{k^s} \sum_{m=1}^{k} \chi(m) \frac{1}{s-1} \sum_{n=0}^{\infty} \frac{1}{n+1} \sum_{j=0}^{n} (-1)^j \binom{n}{j} \left(\frac{n}{k}+j\right)^{1-s}$$

However, as a patch, it works just as well (and in some cases better) than Riemann's original construction:

Hurwitz's Result: $L(s, \chi_n)$ *(the Hurwitz version, not the earlier one) is defined for every single s except possibly $s = 1$.*

Here, by "possibly", we mean that it depends which function $L(s, \chi_n)$ you've chosen. For some choices, $L(s, \chi_n)$ is defined for all s, while for others, it's defined for all s except 1.

---

[1] Minkowski and Hilbert were, of course, known to fans at the time as "Herm and the Hat."

[2] I've been told by several readers that Hurwitz and Piltz never really worked together and only happened to work on the same problem in chronological succession and in fact may never have even met, and also that Piltz was, by all accounts, not a particularly violent man. Shut up, readers. My story's better.

Piltz (center, facing camera) and Hurwitz (left) discuss methods of analytic continuation of L-functions with colleagues.

Despite Hurwitz's extraordinary efforts in discovering these patch functions, we mathematicians have decided to call these things *Dirichlet L-functions* on the grounds that Dirichlet has more syllables and therefore sounds more impressive to say. Hurwitz may not like it, but that's his problem.

10.3 It's Piltzin' Time

Now that we have our functions that are defined for every s and act like the L's that Dirichlet was thinking about, we can finally ask when they hit zero.

This was the question that faced the Adolfs in 1888. Having done his share of the work, Hurwitz now turned things over to Piltz, the brawnier half of the duo, who would attempt to finish the problem using one of his famed finishing moves. Piltz, as you no doubt recall, was not the most adroit when it came to manipulation of these functions, so he did the next best thing; he grabbed $L(s, \chi_n)$, put it in a chokehold, and said, "TELL ME WHERE YOUR ZEROES ARE!"

Unfortunately, Piltz was a bit too strong for his own good, and $L(s, \chi_n)$ was only able to respond, "Mmfghh wmmph thfffff...." before passing out.

This obviously wasn't much help, so Piltz was forced to make an educated guess. He first noticed that some of the zeroes were pretty obvious:

CHAPTER 10. READY FOR RIEMANN!

Piltz's Easy Observation: *Depending on which $L(s,\chi_n)$ you choose, $L(s,\chi_n) = 0$ either when $s = -1, -3, -5, -7, \ldots$ or when $s = -2, -4, -6, -8, \ldots..$*

Figuring out which $L(s,\chi_n)$ corresponds to which of these solutions isn't really hard, so those zeroes are pretty well taken care of.

Now, we get to the harder part: other than the easy answers above, where are these functions zero? Piltz thought long and hard about this problem, looking for all manner of leads and clues and even trying some smelling salts in an attempt to wake $L(s,\chi_n)$ up so he could put it in a chokehold again. Then, he gave up and copied Riemann's answer:

Piltz's Copied Conjecture: *If $L(s,\chi_n) = 0$ and s is not a negative integer then $s = \frac{1}{2} + it$ for some real number t.*

Now, on the one hand, we should be disappointed in Piltz for copying so flagrantly. On the other hand, it looks, miraculously enough, like his conjecture is probably right. In other words, what happens in Riemann's case happens in *every* case, no matter which $L(s,\chi_n)$ you choose; the interesting zeroes are always in the same place.

Piltz stated his new conjecture in 1888 and demanded that it be named "Piltz's Conjecture." However, the mathematical community, although impressed by the simplicity and elegance of the conjecture, were not big fans of Piltz (particularly as many of them had previously been on the wrong end of Piltz's strong-armed tactics), and thus were in no mood to accede to his demands. As a result, the math community decided that the conjecture should instead be called the *Generalized Riemann Hypothesis*[3]:

Generalized Riemann Hypothesis: *Pick an $L(s,\chi_n)$. It doesn't matter which. If $L(s,\chi_n) = 0$ and s is not a negative integer then $s = \frac{1}{2} + it$ for some real number t.*

---

[3]Believe it or not, the grudge against Piltz has continued to this day. Think I'm kidding? According to the American Institute of Mathematics, one of the most important math organizations in the world, "GRH is occasionally called Piltz conjecture, but the conjecture of a Riemann Hypothesis for Dirichlet L-functions is generally viewed as obvious generalization which should not be attributed to a particular person." Yeah. Someone's bitter.

As important as the Riemann Hypothesis is, the generalized version is probably even more important because if it were true, it would imply many, many important results. We'll talk about some of these in the next couple of sections.

10.3.1 Wait, Didn't You Say That The Original Riemann Hypothesis Is The Most Important Conjecture In Mathematics?

Yep.

10.3.2 But Now You Said The Generalized Version Is Even More Important!

Indeed.

10.3.3 Huh?

The reason that these two statements aren't at odds with each other is because mathematicians believe that whatever tools and techniques finally prove the original version of the Riemann Hypothesis will probably give us a lot of insight into how to prove (and may, in fact, actually prove) GRH. This is why the original version is the million dollar problem; it's widely seen as the first domino that will make all of the Riemann Hypotheses fall.

If you're keeping score at home, this means that in addition to making conjectures about mathematics, mathematicians are now also making conjectures about the proofs of these conjectures and how these conjectured proofs of conjectures will help prove other conjectures.

10.3.4 Whoa. I think I need a beer.

Me too.

Chapter 11

GRH Goes To The Races: Daddy Needs a New Pair of Shoes!

In the previous chapter, we mentioned that the Generalized Riemann Hypothesis would immediately prove many other things. In the next two chapters, we demonstrate some of those possibilities; this chapter will focus on a phenomenon known as "prime races," while the next chapter will focus on many of the other problems that would suddenly become much easier if GRH is known to be true.

11.1 Splitting up the World into Threes

For the first of these two chapters, I'm going to start out by splitting the whole numbers into three sets: those that can be written as $3k$ for some k (like 12, which can be written as $3 \cdot 4$), those that can be written as $3k+1$ (like 13, which is $3 \cdot 4 + 1$), and those that can be written as $3k+2$ (like 14, or $3 \cdot 4 + 2$). The split will look something like this:

$3k$:   0, 3, 6, 9, 12, 15, 18, 21, 24, 27, 30, 33, 36...
$3k+1$:   1, 4, 7, 10, 13, 16, 19, 22, 25, 28, 31, 34, 37...
$3k+2$:   2, 5, 8, 11, 14, 17, 20, 23, 26, 29, 32, 35, 38...

Now, as number theorists, we're always looking to turn charts or conjectures or statements about anything into questions about primes, so one might ask, "What happens if I take this chart and remove all of the numbers that aren't primes?" In that case, we have the following:

$3k$:   3
$3k+1$:   7, 13, 19, 31, 37...
$3k+2$:   2, 5, 11, 17, 23, 29...

Do you happen to see a pattern in the $3k$ row? Right, the pattern is, "The list ends abruptly at 3." This makes sense if you think about it; "can be written as $3k$" means "divisible by 3", and prime numbers usually aren't divisible by 3.

Now, what about the other two rows? In those cases, the pattern can be described as, "There are a lot of them." In fact, as you write more and more primes, the pattern will turn out to be, "There are infinitely many in each row."

11.1.1 That's Not A Pattern, Just An Observation

Shut up.

11.1.2 Whatever

Anyway, it is indeed the case that there will be a lot of primes in both the $3k+1$ and $3k+2$ rows. Of course this can only mean one thing...it's time for some racing action!

11.2 Prime Races: Putting the "Math" in "Mathlete"

So here's the deal. Let's take all of the prime numbers from 1-100 and list them in the appropriate category, designating them as $3k+1$ or $3k+2$ as appropriate. Once we hit 100, which category do you think will have more - $3k+1$ or $3k+2$?

Well, let's find out. Make sure you place your bets before the action starts.

$3k+1$:   7, 13, 19, 31, 37, 43, 61, 67, 73, 79, 97

CHAPTER 11. GRH GOES TO THE RACES

$3k+2$: 2, 5, 11, 17, 23, 29, 41, 47, 53, 59, 71, 83, 89

Wow, that was exciting, wasn't it? We number theorists generally refer to these sorts of competitions as *prime races* because deep down, we all like to pretend that we're actually athletes.

Anyway, after the first hundred, the $3k+2$ column is in the lead by a score of 13-11.

What would happen if we expanded the race to 150? Let's take a look:

$3k+1$: 7, 13, 19, 31, 37, 43, 61, 67, 73, 79, 97, 103, 109, 127, 139
$3k+2$: 2, 5, 11, 17, 23, 29, 41, 47, 53, 59, 71, 83, 89, 101, 107, 113, 131, 137, 149

Here, the $3k+2$ column jumps out to a more commanding lead - the score is now 19-15 in favor of the 2's.

What about if we let the primes race all the way up to 1,000? In this case, I have no interest of writing out all of the primes up to 1,000, so you'll have to take my word on this. As it turns out, the $3k+2$ column would continue to lead, but the lead doesn't get much bigger than it was at the 150 mark; the score would be 87-80, giving the 2 column a lead of 7. In fact, if we kept going all the way to 2,000, we would see a similar phenomenon; the $3k+2$ column would lead by a score of 154-146, which is only a difference of 8.

From this, we come to suspect three things:

Three Conjectures (or Suspicions) About Prime Races: Based on what we saw in the last chapter, we will conjecture the following:

1.) *The difference between the two columns is pretty small.* After all, a difference of 8 is pretty insignificant when you're checking all the primes up to 2,000.

2.) *The $3k+2$ column will probably lead most of the time.*

3.) *The point at which I stop bothering to put things in a list is somewhere between 150 and 1,000.*

Are these things true? Well, the third statement certainly is. In fact, I bet the exact answer for the third one is closer to 150 than it is to 1,000. As for the first two, though, those are a bit of a mystery; in fact, right now we don't have particularly good answers for either of them.

On the other hand, if the Generalized Riemann Hypothesis were true, we'd be able to answer both of them in the affirmative:

GRH Beats The Conjectures: *If the Generalized Riemann Hypothesis were true then*

1.) *If we took all the primes up to x and split them into the $3k+1$'s and the $3k+2$'s, the difference in size between the two sets will never be much bigger than $\sqrt{x}$.*[1]

In simpler language, this says that the difference between the two would be pretty small.

2.) *For <u>most</u> x's we can choose, the number of primes up to x in the $3k+2$ category will be bigger than the number of primes in the $3k+1$ category.*

This statement is sometimes called **Chebyshev's Bias**. It is named after Pafnuty Chebyshev, a mathematician who was, for some reason, famously biased against the expression $3k+1$. And Asian people[2].

So, all we have to do is prove GRH and we'd know everything we need to know about these prime races.

It's worth noting that for the second of these two statements (the one cleverly labeled "2") we said *most* and not *all* x's. That's because it is expected that the $3k+1$ column will actually surge ahead on occasion, only to be quickly overtaken again by $3k+2$. In other words, $3k+2$ acts like the mean older sibling who, despite being faster than $3k+1$, occasionally lets $3k+1$ take the lead to keep its hopes up, only to cruelly snatch the lead back

---

[1] The way mathematicians usually express this is to say that the difference will be less than $x^{\frac{1}{2}+\epsilon}$. What they mean by this notation is that the difference will be less than "the square root and a little bit more." In math, for some reason, we've decided to let ϵ, the Greek letter epsilon (the Greek equivalent of "E"), signify "a little bit more." Don't ask me why. None of the words in that phrase begin with E or epsilon or anything close.

[2] Just kidding! I think.

CHAPTER 11. GRH GOES TO THE RACES

shortly thereafter[3].

11.3 Why On Earth Are These Related?

You might be wondering why the Generalized Riemann Hypothesis tells us anything about this sort of question. Well, there's a good reason, and the answer has to do with the fact that we could split our L-functions into products:

Think back to when you were first cordially introduced to the function $L(s, \chi_3)$. At that point, the function looked something like this:

$$L(s, \chi_3) = \frac{1}{1^s} + \frac{-1}{2^s} + \frac{0}{3^s} + \frac{1}{4^s} + \frac{-1}{5^s} + \frac{0}{6^s} + \frac{1}{7^s} + \frac{-1}{8^s} + \frac{0}{9^s} + \frac{1}{10^s} + \frac{-1}{11^s} + \frac{0}{12^s} + \ldots$$

Later on, though, you were introduced to its Batman-like alter-ego,

$$L(s, \chi_3) = \left(\frac{1}{1 - \frac{(-1)}{2^s}}\right) \left(\frac{1}{1 - \frac{0}{3^s}}\right) \left(\frac{1}{1 - \frac{(-1)}{5^s}}\right) \left(\frac{1}{1 - \frac{1}{7^s}}\right) \left(\frac{1}{1 - \frac{(-1)}{11^s}}\right) \left(\frac{1}{1 - \frac{1}{13^s}}\right) \ldots$$

Now, take a second to inspect at the latter version of the function, i.e. the version that appears right above the words you are reading now. You'll notice that some of the primes have a -1 above them, while some of them have a 1. Take a look at which ones are which.

Notice anything? Perhaps something related to the $3k + 1$ and $3k + 2$ business that we've been talking about?

Yep - the primes that have a -1 on top of them are primes (like 2, 5, and 11) that are in the $3k + 2$ column. Similarly, the ones that have a 1 on top (like 7 and 11) are in the $3k + 1$ column. In fact, the only prime that has a zero on top of it is 3, which is also the only prime that appears in neither the $3k + 1$ nor the $3k + 2$ column.

In a nutshell, that's why the Generalized Riemann Hypothesis helps us with this problem. The conjecture that Piltz foisted upon the math community says, in so many words, "$L(s, \chi_n)$ hits zero right where we would

---

[3]For those of you reading this book who happen to be members of my immediate family, let me make it clear that the description of the sibling here is not based on any real persons or events.

expect it to and nowhere else." In the current context, it turns out that this is roughly the same thing as saying, "There's a balance between the primes that have a 1 on top of them and those that have a -1; we don't have many more of one class than we do of the other." If the zeroes are where we expect, the function doesn't do anything weird, and neither do the columns of $3k+1$ and $3k+2$ primes. If the zeroes aren't where you'd expect....well, that would mean that the Generalized Riemann Hypothesis is false, and *that* would be a shock to everybody.

11.4 Moving Beyond 3

Now, the world doesn't have to be split into classes based upon 3; we could have split the numbers using 4 or 7 or 2105 or even 3. Wait. That sentence didn't come out right.

Anyway, let's say we're dealing with 15 because 15, as we all know, is the number after 14.[4] In this case, some classes will have a lot of primes:

$15k+1$:   31, 61, 151,...
$15k+4$:   19, 79, 109,...
$15k+11$:   11, 41, 71, 101,...

Others will have slightly fewer:

$15k+5$:   5
$15k+6$:   None
$15k+9$:   N/A
$15k+10$:   Nope
$15k+12$:   :-(

Just for clarification, let's call the first set of classes (the 1, 4, and 11) *big* or *toweriffic* because they have infinitely many primes, and we'll call the second (the 5, 9, 10, and 12, i.e. the ones that have one or zero primes) *small* or *smurftastic*.

It turns out that we can say the following:

---

[4] 14, of course, is notable for being the number before 15.

CHAPTER 11. GRH GOES TO THE RACES

Thing We Can Say: *Any class that's 15k+something will be either small (i.e. one or zero primes) or towerrific (infinitely many primes). Those are the only options.*

Actually, check that. The same statement would still be true if we replaced 15 with any other number. In other words, for any m we choose, any class $mk + a$ will either be smurftastic or big[5].

Now, the smurfy classes are boring, since they're all either zero or one, so we'll ignore those and focus on the towers instead.

Specifically, what we'd like to know is whether the big classes (the ones with infinitely many primes) will stay close enough together to make for good races. After the excitement of the $3k + 1$ vs. $3k + 2$ races, this mk racing action is obviously bound to draw network attention, and the networks want to know that they're not wasting their time on a race that's not going to be close.

Well, it turns out that a proof of the Generalized Riemann Hypothesis would have some good news for these rather desperate network executives. In particular, we'd know the following:

If GRH Were True: *Let's say you picked a number m to organize around, and two classes, which we'll call a and b. We'll assume that $mk + a$ and $mk + b$ are towered and not stumpy.*

If the Generalized Riemann Hypothesis were true then the number of primes up to x that are in the $mk + a$ column and the number of primes in the $mk + b$ column never differ much more than about $\sqrt{x}$.

---

[5]For those who are interested in the math behind it, it's easy to tell whether a class will be small or big. Notice that in the case of, say, $15k + 12$, both 12 and 15 have a common factor (they're both divisible by 3). We could even factor this expression out: $15k + 12$ is the same as $3(5k + 4)$. But if you have a number that can be written as 3 times something, it's probably not a prime number. So no number that can be written as $15k + 12$ is prime. Likewise 5 and 15 are both divisible by 5; 9 and 15 are both divisible by 3; and 10 and 15 are both divisible by 5. The other classes (1, 4, and 11) don't have any factors in common with 15.

You can probably see where I'm going with this. Pick a number that you're attaching to k; call it m. Pick a number for your class; call it a. If a and m have a common factor (like 10 and 15 did) then there are either one or zero primes that can be written as $mk + a$. If a and m have no common factors besides 1 (like 4 and 15 did) then there will be infinitely many primes that are in the $mk + a$ class.

So it turns out that *every* race, regardless of the m you've chosen, will be as close as the $3k$ race. Given the number of possible choices for m, this is pretty good news for the network executives, who could reasonably fill an infinite amount of airtime with these races. It may not make for the greatest programming in the world, but, hey, it beats watching those 40 year-old reruns of The Dating Game that they show all the time on daytime TV.

Chapter 12

Other Corollaries of the Generalized Riemann Hypothesis

12.1 The Generalized Riemann Hypothesis Implies Everything

In this chapter, I wanted to cover some of the other questions and problems that would be resolved if the Generalized Riemann Hypothesis were found to be true. The list is fairly long and impressive (and pretty varied), so let's dispense with the formalities and the usual beginning-of-the-chapter patter and get right into it:

If we could prove the Riemann Hypothesis (RH) and the Generalized Riemann Hypothesis (GRH), we would be able to prove conclusively that....

12.1.1 ...prime races are exciting.

Whoops - we already did that one. Next...

12.1.2 ...consecutive primes are pretty close together.

In 1850, the aforementioned Pafnuty Chebyshev (or "Nutty", for short, as his friends called him) noticed that the following was true:

Chebyshev's Observation and, Subsequently, Theorem: *Pick a number. Any number (well, any integer bigger than 1). Call it n. Regardless of which n you've chosen, there's a prime number somewhere between n and 2n.*

So if we took $n = 105$ then we're guaranteed a prime number somewhere between 105 and 210. The same thing would happen between 36 and 72 or 1000 and 2000 or 1500000 and 3000000 or whatever n and $2n$ you can think of.

In honor of Chebyshev's impressive work in proving this true, this theorem is usually referred to as *Bertrand's Postulate*.

Wait...

That can't be right...

Oh, yeah. I forgot to mention that Chebyshev, in addition to observing the above, also observed that Joseph Bertrand had noticed the same thing five years prior. Bertrand's work doesn't count, though, because he didn't prove anything. He just observed it and said, "Hey, there it is." Really, that's not that hard. So Nutty gets the credit, even though Bertrand somehow got his name attached.

Interestingly, eccentric Hungarian mathematician Paul Erdős gave a simplified proof of this proposition in 1932. He proved the same thing, but he managed to do it in just four lines[1]. So, now, we're even more sure it's true.

Anyway, the good news is that this statement is correct. It's been proven and reproven and has been given as an exercise to countless graduate students, so we're absolutely positive that this one works. No matter which n and $2n$ you've got, there will always be a prime in between. If you want to be pedantic, you can even say that there's a prime between n and $n + n$, but I'm not sure why you would.

The bad news, though, is that this theorem is probably a little weak. Remember how Nutty promised you that there's a prime between 36 and 72? Well, there are nine of them. There are eighteen primes between 105 and 210, and there are about 100000 or so between 1500000 and 3000000.

---

[1] This fact is even more impressive when you consider that the first line was, "Okay, here goes..."

It's nice of the Nut to promise us one prime, but he probably could have promised a lot more.

The question, then, is how much narrower a range we can take where we can still promise a prime. In other words, we'd like to be able to fill in the following:

Mathematical Mad-Lib: *There is always a prime between n and* $\underset{noun}{\underline{\qquad\qquad}}$.

As is generally the case with Mad-Libs, the answers that we have been able to find to this point have been typically nonsensical, in part because it's hard to answer a deep mathematical question when your only clue is "noun."

On the other, hand, if the Riemann Hypothesis were true, we could eschew all guessing of parts of speech and insert the following:

Mathematical Mad-Lib with RH: *There is always a prime between n and* $\underset{noun}{\underline{n+\sqrt{n}\cdot \ln n}}$.

Noting that $\sqrt{n} \ln n$ is much smaller than n, we actually get a pretty significant improvement here, so that's cool.

It's worth noting that we may actually be able to get a number even smaller than $\sqrt{n} \ln n$ in there. In fact, there are competing conjectures for how low we can actually get, but even getting down to $\sqrt{n} \ln n$ would be a big step.

12.1.3 ...all orthovariant tangifolds are cojective.

Actually, none of these are words. A couple of my friends in grad school used to use these fake words to describe something that was completely incomprehensible, as in,

"Hey Tom, what did the speaker talk about?"
"Cojectivity of orthovariant tangifolds."
"Oh, that sucks."

Hold on, what was I talking about, again? Oh, right, Riemann...

12.1.4 ...several algorithms with widespread use in computer science are guaranteed to run quickly.

Computer scientists are a busy people. In addition to doing things like making the internet and answering the many maddening questions that friends and relatives have about how to fix their computer[2], computer scientists must spend a good portion of their day on important activities such as Starcraft, Warcraft, World of Warcraft, World of Starcraft, Candy Crushcraft, World of Candy Warcraft, and all manner of other vital pursuits.

Because computer scientists' time is so valuable, one of the most oft-asked questions in the field is, "I'm running an algorithm that asks me to input a number and then does some sort of computation with it. Will the algorithm run quickly regardless of the number I plug in? I can't be away from Candy Crushcraft for too long."

Unfortunately, for many algorithms, the answer to this question is often, "Geez, I hope so," which is not a particularly satisfying answer. After all, if I implement a computer algorithm, I'd like to know that it would give me an answer sometime in, say, my lifetime. The difficulty is that how long the computation takes may vary dramatically based on which number you plug in (for instance, it's quite a bit easier to figure out whether 1480120 is a prime number than it is for 1480121 because of that whole "even number" thing, even though the two numbers are really close together), so you may be cruising along, happily putting numbers into your algorithm, and then, BAM!, you plug in the number that takes eons and reduces your computer to a whimpering mess of overloaded circuits and processors and you have to take the computer out back and shoot it to put it out of its misery. This is a scenario that we generally like to avoid if possible.

The hope, then, is that when we make an algorithm, numbers (especially prime numbers) don't conspire together to do anything computationally awful to sink our algorithm. GRH would give us a good feel for exactly when this does and doesn't happen, which would allow us to be more confident in proclaiming whether an algorithm finishes quickly, and, as we all know, confidence is the key to success. Or maybe that's effort. Either way, GRH helps us out.

---

[2]"I ran a large magnet over my computer and now it won't turn on. Which button is the one that undoes that?"

CHAPTER 12. OTHER COROLLARIES OF GRH

GRH may finally allow us to avoid the unpleasantries of computer euthanasia.

Perhaps the most famous example of a test that could benefit from a little Riemannian Magic is the Miller-Rabin Primality Test, sometimes referred to as the Miller-Rabin Prime-O-Matic Prime Tester[3]. Ideally, this test would be able to tell us whether the number you plug in is prime or not, but as of now, it isn't even known whether the test always works. If GRH were true, however, we would know not only that the test would work but that the test would, in fact, smash the world (and Olympic) records for the fastest primality test and change computing as we know it[4].

There are a number of other computer algorithms whose ability to work and/or stop in a reasonable amount of time is currently unknown but would be guaranteed if the Generalized Riemann Hypothesis were true. Such algorithms include the Tonelli-Shanks algorithm to determine whether certain types of equations have solutions, the Lenstra-Pomerance algorithm of factoring large numbers, and, of course, the Notorious BIG algorithm of determining whether Mo' Money does indeed lead to Mo' Problems.

---

[3]Here, by "sometimes", we mean "never."

[4]The current world record is held by a test known as the AKS Primality Test, which was developed in 2002 by Indian mathematicians Manindra Agrawal, Neeraj Kayal, and Nitin Saxena. The Olympic record is held by Michael Phelps.

12.1.5 ...Elvis is dead.

This raises an interesting and important mathematical question that I've never quite been able to answer: how long do we have to wait after a celebrity's disappearance before we can finally put to rest all of the stupid conspiracy theories about how they're actually still alive? I mean, it has to happen eventually, right? I can't imagine people are still claiming that they're the Lindbergh baby. I think we've finally reached that point with Elvis, but it was long, long, *long* overdue.

I bring this up because of all the conspiracy theories I've ever heard, the one about Elvis still being alive might have been the dumbest of all, and, given the level of intelligence generally ascribed to conspiracy theories, that's saying something. Just for fun, let's summarize the arguments made by each of the two sides of the "debate" on whether Elvis was still alive. On the one hand, Elvis was a rock star who ate nothing but fatty foods, popped prescription drugs and amphetamines like Altoids, doubled in size over the last ten years of his life, and had the phrase "failure of the" used in descriptions about most of his vital organs at one time or another. On the other hand, the letters of "Elvis" can be rearranged to spell "Lives." Yep, that's the debate we had for thirty years.

12.1.6 ...decimal expansions of $1/n$ can be long.

When students are first learning how to turn fractions into decimals, one of the most commonly asked questions is, "How much longer do I have to do this?" While this question is usually dismissed as the whiny protestations fifth-graders, there is actually some mathematical interest and validity to it as well[5].

In particular, let's say you're dividing out 1/3. As you might be aware, the decimal expansion for this fraction is extremely simple, and it begins to repeat very quickly:

$$1/3 = .3333333333333333333333.....$$

1/11 is pretty easy as well:

$$1/11 = .090909090909090909090909.....$$

---
[5]Don't tell the fifth-graders, though.

as is 1/101:

$$1/101 = .00990099009900990099...$$

Sometimes, though, a fraction takes longer before it starts repeating. For example, take 1/7:

$$1/7 = .142857142857142857142857...$$

or, worse yet, 1/17:

$$1/17 = .05882352941176470588235294117647...$$

Clearly, these fractions repeat eventually, but it takes them a while to do so (6 terms in the case of 1/7, and 16 terms in the case of 1/17.)

Here's the thing: we know that no matter what n you choose, the decimal expansion for $1/n$ has to start repeating at some point. In fact, if you take any fraction and write it out as a decimal, the decimal has to either end (like .4) or start repeating at some point; that's one of the basic rules of fractions. (A decimal number that never repeats or ends is called *irrational*, and it is known that such numbers can't be written as fractions.) The obvious question, then, is "How long can one of these fractions go before repeating?"

This question has a good answer that was discovered by Carl Friedrich Gauss, the child prodigy we first mentioned in the Riemann Hypothesis chapter. Gauss first published his result in 1801 when he was 24 years old, although, knowing Gauss, he probably first found the answer right around the time he turned 6:

Gauss' Answer: *$1/n$ can only go at most $n - 1$ terms before repeating.*

Note that this says "at most"; sometimes (like when $n = 7$), the decimal goes exactly $n - 1$ terms (in that case, 6 terms) before repeating, while in other instances (like 1/101), the decimal starts repeating long before $n - 1$ terms are reached (in that case, the numbers repeat every 4 terms, whereas $n - 1$ is 100).

Of course, this sort of answer just leads to the obvious follow-up question:

Gauss' Follow-Up Question: *How often will $1/n$ go the full $n - 1$ terms before repeating, and how often will it just give up partway through?*

Number theorists like to change questions that say "how often?" into questions that say, "Does it happen infinitely often?", and Gauss was nothing if not a number theorist, so he followed the script and changed the question:

Gauss Follow-Up Follow-Up Question: *Are there infinitely many n such that $1/n$ goes the full $n-1$ terms before repeating?*

Unfortunately, we have no idea what the answer to this question is. However, we have a good sense of what the answer should be, because proving the Generalized Riemann Hypothesis would actually give us a very specific answer:

Decimal Answer With GRH: *There are infinitely many n for which $1/n$ does the Full Monty and goes $n-1$ terms before repeating. In fact, roughly 37% of prime numbers would have this property.*

This is actually the subcase of a more general conjecture called **Artin's Primitive Root Conjecture**, which sounds like it should be about plants or cavemen or something but is instead about multiplying numbers together. Artin's Conjecture is definitely beyond the scope of this book, but, for what it's worth, GRH would imply the truth of that conjecture, too.

12.1.7   ...Lee Harvey Oswald shot Tupac Shakur with the same magic bullet that killed JFK.

And Sasquatch drove the getaway car.

And finally...

12.1.8   ...OJ Simpson was guilty.

Of course.

The ABC Conjecture

Chapter 13

ABC: What the Alphabet Looks Like When D Through Z are Eliminated[1]

13.1 The Loud Beginnings of a Beautiful Conjecture

Of all of the conjectures in this book, the ABC Conjecture is by far the least historic.

Unlike 150-year-old Riemann Hypothesis or the Twin Prime Conjecture whose age is measured in millennia, the ABC Conjecture did not exist until as recently as 1985. Of course, a conjecture that's been open since 1985 is still a rather impressive proposition, as such a conjecture can then be said to have stumped mathematicians for almost 30 years; nevertheless, the ABC Conjecture is younger than I am, which means that I can't really wax poetic about how the world was "a very different place when this conjecture was formed" or other such nonsense (like I did in some of the other chapters) unless I want to make myself feel really, really old. So that's kind of disappointing.

However, whatever gravitas the conjecture lacks because of its youthful age is no match for the amount of gravitas it lacks in how it was discovered. Yes, the ABC Conjecture is, to date, the only major math conjecture known

---

[1]This is a variant of an old joke by former comedian Mitch Hedberg. RIP, Mitch

CHAPTER 13. ABC: THE ALPHABET WITHOUT D THROUGH Z

to have been discovered at a cocktail party.

You see, mathematicians David Masser and Joseph Oesterle were at a cocktail party[2], surrounded by boisterous partygoers[3], when they began to discuss a recent paper that had written the year prior about polynomials[4]. Masser and Oesterle started to think about what would happen if you took the paper and replaced every mention of "polynomial" with "integer." This turned out to be harder than expected, largely because "Wake Me Up Before You Go-Go" by Wham! had just come on the radio and several of the people around Masser and Oesterle were singing along at the top of their lungs, but the mathematical pair soldiered through. Finally, after a bit of discussion and a lot of thought, it was Masser who saw the light and realized what it was the duo could accomplish with this simple idea; in one of the memorable moments in math history, Masser then turned to his colleague said, "Hey Oesterle! We can....what the....?"

CRASH! Just then, a dancing partygoer came tumbling through the table on which Masser and Oesterle were working.

After cleaning up a bit, Masser clarified that what he was going to say was, "Using the ideas from the paper, we should come up with a simple proposition about integers that would change the way we look at additive relations between integers and completely revolutionize the study of Diophantine equations. Also....oh c'mon, not again...," CRASH - and it was back to the broom and paper towels.

Once round two of cleanup had been done, Masser and Oesterle set to work. They realized that Masser's idea was a good one, and so the pair sat down and hammered out a stunning conjecture about the relations between numbers in equations - relations that, despite their fundamental nature, had somehow eluded mathematicians since the dawn of time. The equations that the two mathematicians investigated were so simple that it was clear that there would be immediate applications throughout all of mathematics.

Then Masser and Oesterle made a solemn vow never to do important work at a raging party again.

---

[2]This part is actually true

[3]This part might be slightly less true.

[4]Yeah, we mathematicians know how to party.

13.2 Hold On - How Simple are We Talking Here?

The conjecture applies to nearly every equation of the form

$$a + b = c.$$

That's why it's called the ABC Conjecture.

13.3 Oh. Yeah, That's Pretty Simple.

See what I mean? That equation is bound to show up in quite a few places.

13.4 Wait, Wait - Hold On a Minute. How Is It Even Possible That They Found Something New To Say About $a + b = c$?

It's kind of amazing when you think about it. $a + b = c$ is an equation that we've studied almost continuously since the discovery of algebra in the 9th century; saying "We found something new about $a+b = c$," seems kind of like saying, "We found more elements in water besides hydrogen and oxygen," or "We were wrong about the number of planets in the solar system[5]." The whole concept seems, on its face, insane.

The insight that Masser and Oesterle had, though, was that instead of looking at the numbers a, b, and c themselves, we could look at the *factors* of a, b, and c, and see if there were any rules that we could form about them. Doing this is a bit counter-intuitive because $a+b = c$ is an equation that talks about addition and mentioning "factors" usually means that we're talking about multiplication, so what this is saying is basically, "Let's see if there are any unexpected relationships between addition and multiplication."

---

[5] Actually, this second one might be a bad example

13.5 Hmmm. That's Not Any Less Mind-Blowing.

You're right. It's probably best if we stop talking in vague generalities and just move on to the actual math.

Chapter 14

Radical! Mason and Oesterle's Excellent Adventure

Before we get to the actual statement of the ABC Conjecture, we have to define a new function that will allow us to express the key ideas of the conjecture. To do so, we must first return to the world of prime numbers....

14.1 Primes? Primes!

As you may have gathered from pretty much every word written in every chapter before this one, number theorists *love* primes. Like, really love primes, to the point where a restraining order might be required. That's just how we roll. So when we see an integer just sitting there like this:

$$600$$

our first instinct is to say, "Let's split the integer into prime factors!":

$$600 = 2 \cdot 2 \cdot 2 \cdot 3 \cdot 5 \cdot 5,$$

or, better yet,

$$600 = 2^3 \cdot 3 \cdot 5^2.$$

You'd think that this would be enough prime-related carnage to satisfy our compulsion. Sometimes, it is. Other times, however, we might say, "No one cares about these exponents. All we want are primes! PRIMES!" In fact, we even have a function, called the radical (denoted $rad(x)$, so named because

"Radical!" was considered a cool thing to say when the conjecture was made in 1985), where we take the prime factorization and strip the exponents:

$$rad(600) = 2^{\cancel{3}} \cdot 3 \cdot 5^{\cancel{2}} = 2 \cdot 3 \cdot 5.$$

Note that $2 \cdot 3 \cdot 5 = 30$, so we could write

$$rad(600) = 30$$

to get the same info across.

Want another example? Sure you do! Here's one:

$$112 = 2^4 \cdot 7,$$

so

$$rad(112) = 2^{\cancel{4}} \cdot 7 = 14.$$

That was awesome[1].

Note that the radical of a number is quite often much smaller than the original number. Of course, this isn't always the case:

$$rad(30) = 2 \cdot 3 \cdot 5 = 30,$$

$$rad(17) = 17,$$

but it's true often enough to be a useful tool in simplifying things considerably. In fact, the ABC Conjecture is based upon this idea:

General Question: *In general, how much smaller is the radical of a number than the number itself?*

Are most numbers going to be unaffected (like 30)? Are they going to be significantly smaller, like 112? Are they somewhere in between? Why is this function considered so "radical," anyway? And why do we use "radical" to mean "awesome!"? And what about the phrase "That's radicool!"? Wasn't

---

[1] If you're wondering why we'd bother creating such a function, the mathematical idea here is that there are times where we only care *which* primes divide a number (i.e. does 7 divide our number or doesn't it?); in these scenarios, the actual number of times that each prime divides into the number isn't nearly as important.

CHAPTER 14. RADICAL! EXCELLENT!

that a stupid expression? Who came up with that one? And why? Oh, and remember "Don't have a cow, man!"? What ever happened to that phrase?

These are all very important questions, so a conjecture that merely addressed these questions would have been enough for most people. However, Masser and Oesterle are not most people. They realized that they could come up with a way to merge these questions with other ideas in number theory and make an even more ambitious and sweeping conjecture with broad implications across mathematics. Like midgets at a shooting range for really tall people, Masser and Oesterle decided that it was time to aim high....

Chapter 15

Towards A Meaningful ABC

15.1 Introducing B and C into the ABC

When Masser and Oesterle stumbled upon these questions, they realized that it was a golden opportunity to say something interesting about both the radical function *and* equations themselves.

To explain, let's go back to the equation we discussed earlier:

$$a + b = c.$$

Here, I'm going to require that a, b, and c all be positive integers so that we don't have to deal with negatives or zeroes or fractions or imaginary numbers or radicals or whatever else because, seriously, the hell with that crap. The numbers here will be whole, they will be positive, and they will be fantastic.

In fact, I'm going to assume that a, b, and c don't have any common factors (besides 1). I'll explain why in a later footnote, but for now I'll just say that having numbers in an equation with common factors is like the mathematical version of inbreeding; it's disgusting, and everything that comes out is dumber because of it. And that's all I'll say about that.

Now that we've gotten those unpleasantries out of the way, we have the following question:

New, Improved ABC Question: *Take the equation from above:*

$$a + b = c,$$

where a, b, and c don't have any common factors. How do a, b, and c compare to rad(abc)?

Now, why is this formulation an improvement, apart from the fact that we've eliminated the unsavory practice of inbreeding in mathematics? Well, it turns out that this is a really, really clever way to take an average; although it's entirely possible that one of the numbers in the equation might be something like 2^{1320} that gets a pretty big haircut from the radical function, it's pretty unlikely that all three of the numbers are going to be that way. Basically, it's the number theory version of taking a random poll; 2^{1320} might have a strong opinion on the radical function, but having two other numbers in the sample help counterbalance 2^{1320} and give a more reasoned answer[1].

Now, you might be wondering, "In that case, why did you only include three numbers? Most polls have 500 numbers in their sample. Why don't you set up an equation with more variables?"

If you posed this question to a number theorist, he or she would probably tell you that three turns out to be enough variables to say something interesting; in fact, he or she would say, having 500 variables would just be overkill because the result probably wouldn't get that much better. But really, we're just lazy. So...three variables it is.

What's even better about this method, though, is that it takes the radical and somehow relates it to equations. This means that we can now apply all of the stuff we know about primes and radicals to cases where we're trying to solve equations. As you might be aware, mathematics has many equations, so this will help out quite a bit.

---

[1] Now you can see why we eliminated numbers that have common factors; they would all be affected similarly by the radical function. For instance, if we took a, b, and c to be something like
$$2^{1320} \cdot 3 + 2^{1320} \cdot 4 = 2^{1320} \cdot 7,$$
you can see that all three of the numbers would be decimated by the radical function.

This would be the mathematical equivalent of taking a poll on presidential approval and choosing all of your participants from Barack Obama's immediate family.

15.2 The Almost-Conjecture

So, how large of an effect can the radical function have on a, b, and c simultaneously?

Well, it's still going to depend to some degree on which a, b, and c you choose. For instance, if you pick the equation

$$14 + 17 = 31,$$

you would notice that

$$rad(14 \cdot 17 \cdot 31) = 2 \cdot 7 \cdot 17 \cdot 31 = 14 \cdot 17 \cdot 31,$$

which is to say that the radical function doesn't really do much to this triple of numbers. On the other hand, if you pick

$$25 + 144 = 169,$$

we have

$$rad(25 \cdot 144 \cdot 169) = rad(5^2 \cdot (2^4 \cdot 3^2) \cdot 13^2) = 5 \cdot 2 \cdot 3 \cdot 13 = 390.$$

Since $25 \cdot 144 \cdot 169 = 60400$, the radical function has a pretty significant impact on these numbers.

What we've eliminated, though, is the case where the radical function is *absurdly* small relative to the three numbers chosen. In fact, for most choices of a, b, and c, we can say something like this:

Thing That Is True The Vast Majority of the Time: *If $a + b = c$ then the following isn't always true but is true the vast majority of the time:*

$$c < rad(abc).$$

Now, this may not seem terribly restrictive (after all, even when we had $25 + 144 = 169$, the radical of $25 \cdot 144 \cdot 169$ was nowhere near as small as 169). However, you should keep in mind that we're looking here for rules that will satisfy *every equation in the world*, so even a seemingly small discovery like this would have huge effects throughout mathematics.

Unfortunately, as you may have noticed from the chosen verbiage for the bolded words above, the statement "$c < rad(abc)$" doesn't *quite* work all the time; indeed, there are still the occasional a, b, and c where the radical manages to sneak just a tad below c. For example:

$$243 + 100 = 343$$

gives

$$rad(243 \cdot 100 \cdot 343) = rad((3^5) \cdot (2^2 \cdot 5^2) \cdot (7^3)) = 3 \cdot 2 \cdot 5 \cdot 7 = 210,$$

and 210 is certainly less than 343.

The good news is that such cases are fairly rare, and most triples you can pick are decent enough to keep their radicals up at respectable levels. The bad news is that a statement that is only true most of the time doesn't work as a conjecture; after all, "true most of the time" is a synonym for "false some of the time," and if we want to come up with a conjecture that will change the field of mathematics, it would help if said conjecture were, you know, not false.

15.3 The Almost Conjecture, But With Less "Almost" and More "Conjecture"

So, how do we fix our rule and make it true *all* of the time?

Well, one way is to raise the right hand side to a power:

$$c < rad(abc)^{102562781}$$

No, no, a smaller power than that.

$$c < rad(abc)^2$$

Yeah, that's better. In fact, this is part of the ABC Conjecture:

Part of the ABC Conjecture: *Let a, b, and c be positive integers that don't do anything creepy like possess common factors. Oh, and $a + b = c$. If all of that stuff is true then*

$$c < rad(abc)^2.$$

Now, the good news is that this is probably correct (which is why we called it a conjecture instead of something like "Bad Guess"). For instance, in the delinquent case I gave before with $243 + 100 = 343$, we had that the radical was 210, and $210^2 = 44100$ is definitely bigger than 343.

The only drawback here is that squaring $rad(abc)$ is probably overkill, akin to washing the dishes with a fire hose. Squaring the right-hand side is a powerful operation that makes the term much, much bigger (after all, we said that an exponent of 1 would work for almost all a, b, and c, and even in the truant case discussed in the previous paragraph, 44100 is *way* bigger than 343), and we'd like to know if we can get something smaller than 2 in the exponent.

Well, can we? For instance, let's say we wanted to lower the exponent from 2 to 1.5. Could we do that?

Um, sort of...

Another Part of the ABC Conjecture: *Let a, b, and c be good like before.* IF C IS SUFFICIENTLY LARGE *then*

$$c < rad(abc)^{1.5}.$$

Did you happen to notice that I slipped in a few words there? I tried to be subtle, but my caps lock key got stuck. Oh well.

Anyway, the new words say that this isn't true for *every* a, b, and c, but once c passes a certain threshold, it'll always be true.

It's worth noting that we have absolutely no idea where this threshold is. We know that it's pretty large (we've computed some pretty big numbers where the equation doesn't yet work), but as far as how large we're talking.....no clue. However, even knowing that such a threshold exists would be a big deal; one of the difficulties of math is that we're always trying to show that something is true for every possible number (all the way out to infinity), so if we knew that there was a (finite) threshold where we could stop checking things, that would seriously cut down on the work we have to do and would therefore be awesome.

15.4 Movin' on Down: What The Jefferson's Theme Song Would Be if You Turned The TV Upside Down[4]

---

[4]Viewer discretion is advised.

CHAPTER 15. TOWARDS A MEANINGFUL ABC

Now that we've got this adjustment where we say "for sufficiently large c" and everything is magically better, you might be wondering how much further we can get this exponent to go.

Well, the bad news is that we still can't get it all the way down to 1 - no matter how high the threshold, you can always find a non-compliant a, b, and c where c is bigger than your threshold. The good news, though, is that we can get really, really close to 1. In fact, we should be able to get about as close to 1 as you want:

The Full ABC Conjecture: *Let a, b, and c be good like before, and let r be a positive number that's as close to 0 as you like (without being 0). If c is sufficiently large then*
$$c < rad(abc)^{1+r}.$$

Of course, what we mean by "sufficiently large" will depend on which r we've chosen; the threshold we pick for $r = 0.5$ will be way too small for $r = 0.05$, and what works for $r = 0.05$ probably won't work for $r = 0.0001$. In effect, the ABC Conjecture is not one conjecture but tons and tons of mini-conjectures; for each possible r, you have to both prove the conjecture and find the threshold for which it works. That said, we don't have to get them all at once; if we could prove this for *any* choice of r, it would be a huge advance in mathematics and would give us a whole lot of new information about all kinds of other problems.

Chapter 16

Current State Of Affairs: Where We Are and Things That The ABC Conjecture Would Prove

16.1 How Close Are We to Proving the ABC?

Good question. Remember the statement that for sufficiently large c,

$$c < rad(abc)^{1+r}?$$

Well, when I first started writing this book, most of the results we had gotten looked more like this:

$$c < e^{rad(abc)^{1+r}}.$$

That's not very good. Any time you're raising e to a power, you're making things a whole lot bigger.

The whole situation changed rather abruptly in 2012, though, when Japanese mathematician Shinichi Mochizuki shocked the world with the announcement of a possible proof of the ABC Conjecture. The proof was undoubtedly a novel one; using methods from a field called "inter-universal Teichmuller geometry," Mochizuki had discovered a completely new attack to the problem that no one had ever considered, and he put these ideas to work in a 500-page opus of mathematics that (in his estimation, anyway) resolved the problem once and for all.

Upon hearing this announcement, most mathematicians had exactly the same response that you probably did when reading the previous paragraph; they proclaimed, "What the hell is inter-universal Teichmuller geometry?" This wasn't one of those instances where mathematicians were merely surprised by a new result; Mochizuki had been working independently on his own ideas for over 10 years, and no one had the slightest idea what he was doing. In fact, even the name of the field ("inter-universal Teichmuller geometry") was a Mochizuki invention that very few mathematicians not named Mochizuki had ever heard before the pronouncement.

As a result, the question of whether or not the proof is correct is still (as of this writing) a very open question. There aren't many mathematicians who have the time or inclination to learn a whole new area of mathematics and wade through 500 dense pages of unfamiliar terms and ideas for a proof that may or may not even be correct; it's very slow work, and learning a bunch of new ideas that turn out to be wrong has the potential to be a rather significant waste of one's time. I can report that there have at least been some preliminary attempts to parse this jungle of mathematics by a couple of enterprising young mathematicians who are clearly either more ambitious or more masochistic than I am, but it may be quite a while before we know whether the ABC Conjecture has actually been resolved.

Anyway, there's not an awful lot more we can say about this purported proof, so let's move on to the better-understood question of why we care so much about the ABC Conjecture:

16.2 Things That The ABC Conjecture Would Prove

Much like most of the other conjectures in this book, a proof of the ABC Conjecture would have quite a few effects in other places in number theory. Examples include:

1.) **Fermat's Last Theorem.** For many years, Fermat's Last Theorem was one of the most popular unsolved problems in all of mathematics. I mention it in several other places in this book, but it bears stating in some

CHAPTER 16. CURRENT AFFAIRS

detail here because a.) it's important, and b.) it's one of the most obvious motivations for studying the ABC Conjecture.

As the story goes, when mathematicians were studying Fermat's notes and works after he died in 1665, they found a curious comment scribbled in the margin of Fermat's copy of Diophantus' *Arithmetica*:

$x^n + y^n = z^n$ has no solutions if $n > 2$

and, underneath that,

I have a truly marvelous proof of this proposition which this margin is too narrow to contain.

This simple statement set off a three century search for a proof of this proposition, as the conjecture that would go on to be called Fermat's Last Theorem stumped many of the greatest minds in the history of number theory. It was finally solved in 1994 by Andrew Wiles, who wrote a 192 page proof of the theorem, then discovered that the 192 pages weren't quite enough to actually prove the theorem, and finally released an additional 12-page paper that completed the proof. Wiles' proof depended on a method known as "the kitchen sink," where he pretty much took everything in modern mathematics that wasn't bolted down and threw it at the problem and eventually got things to work. It's a very brute force-ish proof - effective, but not efficient by any means[1].

On the other hand, if the ABC were true, Fermat's Last Theorem could be easily proven in less than a page. That's right, I said *less than a page*. It's that powerful.

Other results include....

2.) **Progress on Wieferich's Primes**. Yeah, there's that. And

---

[1]Incidentally, one of the great questions in math history is, "Did Fermat actually have a solution to this problem?" Obviously, the method of proof that Andrew Wiles used was not available to Fermat at the time, so the question is really whether Fermat had a simple, low-tech proof of this problem. I'm firmly in the "No" camp for two reasons: a.) Fermat is known to have stated theorems that he couldn't actually prove (we know this because he occasionally stated "theorems" that turned out to be false), and b.) if there actually were a simple solution, Euler probably would have figured it out.

CHAPTER 16. CURRENT AFFAIRS

3.) **Szpiro's Conjecture About Elliptic Curves.** Is a nice one too. And

4.) **A Result about Class Numbers** is something I would bother explaining, except that....

5.) **Seriously, FERMAT'S BLEEPING LAST THEOREM.** I mean, I don't even know why I should bother with the other results. A proof of the ABC Conjecture turns the most difficult problem of the last 300 years into an undergraduate classroom exercise. I think I should be able to just stop here.

In fact, I will[2].

---

[2] A full list of implications of the ABC Conjecture is available at

http://www.math.unicaen.fr/~nitaj/abc.html

Chapter 17

Appendix B: Why ABC gives us Fermat's Last Theorem

Earlier, I made a statement that ABC would be able to prove Fermat's Last Theorem very, very easily. You might remember it - it was a statement laced with expletives and stuff. Anyway, the more mathematically inclined among you might have said, "That's an awfully bold claim! I'd like to see this purported proof." Actually, the less mathematically inclined among you might have said that, too. I don't know. I'm not really in the business of predicting what people will say.

Anyway, in this section, I wanted to sketch out how this proof would go. If you're afraid of math, no worries - the proof is obviously pretty short and doesn't require any advanced math. Honestly, that's probably a good description of what's most amazing about this proof - it's short and doesn't require advanced math.

All right, let's do this...

17.1 The First Step: State the Facts

To start out, we're going to need two facts, which I will cleverly label Fact 1 and Fact 2:

Fact 1: The radical of x^n is the same as the radical of x, which makes sense since the radical gets rid of exponents. If we wanted to write this in math, we would say $rad(x^n) = rad(x)$.

Fact 2: $rad(x)$ can't be bigger than x. Of course this is true - the radical function strips exponents, so it either makes your number smaller or keeps it the same. In math language, we would write $rad(x) \leq x$.

Now, let's say we had a solution to our Fermat equation

$$x^n + y^n = z^n.$$

(We'll assume that x, y, and z are all positive.) Then we can add Fact 3:

Fact 3: x and y are both less than z. You add two positive numbers, you get a bigger positive number. That one makes sense.

17.2 Bring on the Conjecture!

Next, it's time to bring in our ABC Conjecture. Remember that the ABC Conjecture said something like

$$c < rad(abc)^2.$$

There were all different variants, but we'll go with this one because it's a nice, round exponent and I don't feel like dealing with decimals.

Let's apply the ABC Conjecture to our Fermat equation, which I'll repeat here:
$$x^n + y^n = z^n.$$
The ABC Conjecture would then say that

$$z^n < rad(x^n y^n z^n)^2.$$

Now, we have the following observations:

1.) $rad(x^n y^n z^n)^2$ is the same thing as $rad(xyz)^2$ because of Fact 1 (the radical takes out exponents, so we might as well help it out).
2.) $rad(xyz)^2 \leq (xyz)^2$ because of Fact 2.
3.) x and y are both less than z (Fact 3). So $(x \cdot y \cdot z)^2 \leq (z \cdot z \cdot z)^2$.
4.) $(z \cdot z \cdot z)^2$ is the same thing as z^6.

5.) Running by the pool is prohibited.

Putting all of these together, we have that

$$z^n < z^6.$$

That's pretty restrictive. Since z is positive (and an integer), what this really means is that n has to be less than 6.

In effect, we've managed to reduce Fermat's Last Theorem down to some very easy cases, since the cases where n is less than 6 aren't all that difficult and were solved by the early 1800's.

Impressive, no?

The Birch-Swinnerton-Dyer Conjecture

Chapter 18

Preface

When I started writing this chapter several years ago, it was supposed to be a manuscript about Fermat's Last Theorem.

Unfortunately, as some of you may know, the proof of Fermat's Last Theorem has to be the most over-exposed mathematical development in 20th century mathematics, so it was tough to say anything new. Granted, "the most over-exposed mathematical development in 20th century mathematics" is a pretty dubious distinction, akin to being the best singer at a karaoke contest for the deaf, but it was still a pretty big deal at the time. After Andrew Wiles announced his proof in 1992 (and the fixed version of the proof in 1994), a cottage industry sprung up around FLT, resulting in an onslaught of technical books, non-technical books, bestselling books, documentaries, expositions, limericks, artistic renderings, and t-shirts. References to the proof permeated TV shows, movies, and songs. The Gap offered to make a jeans commercial with Andrew Wiles. There was even a musical written about Fermat's Last Theorem. I swear to God. A musical about number theory.

In short, the point is that there's no real angle left for me to take for FLT.

By contrast, the Birch-Swinnerton-Dyer Conjecture is far more approachable. Mathematicians understand its importance but can't communicate it to the outside observer. Outside observers have never heard of it. And most importantly, there's a million dollar reward attached to it, which gets everyone's attention immediately.

Plus, I can take the section I wrote about elliptic curves (which are a key part of the proof of FLT) and use it in the writeup of BSD instead.

CHAPTER 18. PREFACE

So.....BSD it is.

I'll talk about it in two parts. First, I'll discuss the weak version of BSD; this will give the reader a sense of what it is we're doing with these functions and curves. After that, I'll get into the full-strength version of the conjecture, which is 47% stronger and 38% more effective at fighting cavities. It'll be a good time for all.

Chapter 19

Elliptic Curves: Nothing to Do With Ellipses

19.1 Minimal Work, Maximal Money

One of the great questions in mathematics has always been, "What's the easiest thing that we can do that still qualifies as work[1]?" For example, let's say we had an equation like

$$y = x + 2.$$

Then we might be tempted to ask a question like, "What are possible solutions for x and y?" Unfortunately, that's way too broad of a question to answer particularly well, so we probably want to restrict ourselves by asking something like, "When are x and y both rational numbers[2]?" In other words, we'd like to know the following: which fractions (rational numbers) could we plug in for x and y to make this equation true?

Well, hey, I can find a bunch of answers: $x = 2$ and $y = 4$, $x = \frac{1}{2}$ and $y = \frac{5}{2}$, or $x = -1$ and $y = 1$, just to name three. There are a bunch more, too, and they're all really easy to find. Wait, I just found another one: $x = \frac{3}{10}$ and $y = \frac{23}{10}$. Oh, and there's always $x = 0$, $y = 2$. I could do this all day.

So, is this question too easy to qualify as work? Are you kidding? Of

---

[1]On second thought, I'm not really convinced that this question is unique to mathematics.

[2]If you've forgotten this vocabulary from the introduction, rational numbers are fractions (like $\frac{5}{6}$) where the top and the bottom are both integers. Note that the term "rational numbers" includes whole numbers like 3, since 3 can be rewritten as $\frac{3}{1}$.

CHAPTER 19. ELLIPTIC CURVES

course it is. It's way, way too easy; in fact, I pretty much just answered it. There's no way we can get funding for a question that I can answer in half a page. We'll have to rachet up the difficulty a bit if we want to get paid, especially if we want to get in on high-paying NSF government grants.

What about if we had an x^2 on the right? Something like:

$$y = x^2 + x + 6?$$

Looks like it might be harder, but it's still too easy[3]. No money there.

Maybe if we had a y^2:

$$y^2 = x^2 + x + 6?$$

Turns out that's still too easy. Since both x and y are raised to the same power, the problem turns out to be no challenge at all. You can pretty much just take the square root of both sides and kiss your grants goodbye.

How about if we had an x^3? Maybe something like:

$$y^2 = x^3 + x^2 + x + 6?$$

Bingo. You've hit paydirt. This is the easiest equation (in terms of powers of x and y) for which things actually get interesting.

Now, let's say you had the equation above. Then, according to the script outlined above, you would say, "When are x and y both rational numbers?"

And I, motivated by the fundamental importance of such a question and its implications on the advancement of mathematical knowledge, and, indeed, upon human knowledge itself, would respond, "Good question. Let's name everything and then apply for some grants! Yay money!"

The first thing we name is this function. In general, if we have a function with y^2 on one side and x^3 on the other, we have what is called an *elliptic curve*. To be more specific, we have the following definition:

Definition: *Say we have a function that looks like*

$$y^2 = x^3 + Ax^2 + Bx + C$$

---

[3]For those of you who remember the quadratic formula, that formula basically solves this sort of problem completely. For those of you who don't remember the quadratic formula, you have no idea what I'm talking about in this footnote.

CHAPTER 19. ELLIPTIC CURVES

except that A, B, and C have mysteriously been replaced by integers. We call this an **elliptic curve**.

Often, we call this curve "E," especially when talking about it behind its back.

You may be wondering why they call these things elliptic curves and what they have to do with ellipses. The answers, as it turns out, are "I don't know" and "Nothing." Hope that helps[4]!

The second thing we name is the whole thing about x and y both being rational. If x and y are both rational, we call x and y a *rational point*. So instead of saying, "When are x and y both rational numbers?", we can say, "Where are the rational points?" For example, let's go back to our moneymaker that we found earlier:

$$y^2 = x^3 + x^2 + x + 6?$$

Let's say we notice that $x = 1$ and $y = 3$ satisfy this equation[5]. Then we would say that $x = 1$, $y = 3$ is a rational point; if we're pressed for time, we might write that $(x, y) = (1, 3)$ is a rational point instead (although it means the same thing).

Thus, we can restate what we were asking earlier:

Elliptic Curve Question: *Say we have an elliptic curve, which we've named "Eazy E", or "E" for short. Does E have rational points? Inquiring minds want to know*[6].

This looks like a very good question. Well, it isn't. Or rather, it isn't yet. It'll need a little bit more tweaking before it becomes something worth pursuing.

---

[4]It is claimed that elliptic curves were given their name as a result of their connection to elliptic integrals. Yeah. Sure.

[5]They do, after all. Just plug in 1 for x and 3 for y and you get that the left side equals the right. What, do I have to do everything around here?

[6]True story that I might have just made up: '90s Rapper Eazy E chose his rap name as a tribute to mathematician Leonhard "Eazy" Euler, father of elliptic curves.

CHAPTER 19. ELLIPTIC CURVES 95

Leonhard Euler (right) and protege Eazy-E.

19.2 Turning the Ugly Question into A Beautiful Butterfly. Or Swan. I Forget Which.

So why isn't the Elliptic Curve Question above a good question? To answer that, we should probably introduce a little bit about the background of elliptic curves, so let's do that.

It was actually the great Leonhard Euler (shown above) who began the study of elliptic curves because, as we all know, Euler was all about the Benjamins[7]. Elliptic curves turn out to be a mathematical object that show up in all sorts of places, from classical applications in geometry to really modern applications to cryptology and computer encryption schemes. They even play a large role in the proof of Fermat's Last Theorem that is mentioned in the preface. In short, they're everywhere.

Euler realized that the resolution to the Elliptic Curve Question above is not that hard; it's often easy to find a rational point, or two rational points, or even three rational points if you're feeling adventurous. It turns out that the interesting distinction, then, is not whether a curve has *any* rational

---
[7]Benjamin Franklin, who was a contemporary of Euler's, found the obsession creepy.

points but whether a curve has *infinitely many* rational points.

For instance, I mentioned that elliptic curves are the basis for some types of cryptographical systems. Well, the more rational points your curve has, the more ways you can set up your system. Which do you think makes for a system that's harder to crack: a system based on a curve with infinitely many rational points (which would give you infinitely many choices for how you can set up the system) or a system based on a curve with three rational points? I don't know about you, but I'd go with the "infinitely many" option[8].

As a result, Euler decided that the "infinitely many rational points?" question is the one we should be asking about an elliptic curve. If the number of points is infinite, we have found what we are looking for, and we will party long into the night. If they are finite, we curse and mutter and lose our NSF grant. The stakes are high.

Bigger Elliptic Curve Question: *Let's get back to our elliptic curve E. Does E have infinitely many rational points? Or are there only finitely many, like so many failed curves before?*

Answer: *Well, it's going to depend on which curve you've chosen.*

And thus, we have our problem. We'd like a quick way to find the answer to this question. We don't have it, so we're forced to the usual tactics of laborious computation, tiresome computer searches, bizarre seances with witch doctors, and whatever else we can throw at the problem. These methods waste not just time and computing power (since computer searches can be extremely inefficient) but also money (since most witch doctors are not American and are therefore not covered under NSF grants). Our hope is that we can find a simpler, more computationally effective method of finding our information about elliptic curves without all of the hassle, consternation, or mysticism that the current methods require.

---

[8]If you would choose the "three rational points" or "no rational points" option, please let me know if you ever run a bank or other major company so that my computer scientist friends and I can do some "research" into your "fundamentals."

Chapter 20

Modular Arithmetic: Why Telling Time Actually Counts as Doing Math

In order to talk about our new ideas for finding information about elliptic curves, we have to introduce a new form of arithmetic called "modular arithmetic." It sounds fancy to say, "new form of arithmetic," but, really, this is just one of those turns of phrase that we mathematicians say in order to make people think that we're way smarter than we actually are. In reality, I'm just teaching you how to read a clock.

20.1 Clock Arithmetic: Number Theory As Taught By Your Watch

All right, so imagine you had a clock[1]. Or else look at the one you have. Right now, it's about 3 o'clock[2]. In 6 hours, what time will it be? You guessed it: 9 o'clock.

$$3 + 6 = 9.$$

Say it were now 9 o'clock. In 6 hours, what time would it be? You guessed it: 15 o'clock.

---

[1] Just for clarity's sake, we're using a 12-hour clock here.
[2] If it's not 3 o'clock, you should assume that you are reading this chapter in the wrong time zone. Shame on you.

CHAPTER 20. MODULAR ARITHMETIC: TELLING TIME

$$9 + 6 = 15.$$

Wait, that can't be right. My clock only goes up to 12 and then goes back to 1. In 6 hours, then, it would be 3 o'clock.

This defines a new kind of addition; we can call it "clock arithmetic," though number theorists often call it "modular arithmetic." To indicate that this is modular arithmetic instead of regular arithmetic, we often write "≡" instead of "=", and we write (mod 12) to indicate that twelve is where the clock resets. Thus,

$$3 + 6 \equiv 9 \pmod{12},$$

but

$$9 + 6 \equiv 3 \pmod{12}.$$

By convention, instead of 12, we say 0, so picture a clock where the 12 has been replaced by a 0 and you're good:

$$4 + 8 \equiv 0 \pmod{12}.$$

This works for subtraction as well. For example, if we start at 2 o'clock and go back 4 hours, we get not -2 o'clock but instead 10 o'clock:

$$2 - 4 \equiv 10 \pmod{12}.$$

Multiplication strays from the analogy a little bit, but it still works: if we start at 0 and go forward 3*10(=30) hours, we end up at 6 o'clock, so

$$3 * 10 \equiv 6 \pmod{12}.$$

You know what's awesome about this? This "mod 12" thing means we only have to deal with the numbers from 0 to 11; since we'll never have, say, a 22 o'clock, all the other numbers don't matter.

What if we lived in a country[3] where clocks were 7 hours instead of 12? A clock would go from 0 to 6, and after 6, it would go back to 0 and start again. Here, our clock arithmetic would look a little different. Let's say it's 3 o'clock now. In 1 hour, it would be four o'clock:

$$3 + 1 \equiv 4 \pmod{7}.$$

---
[3] No, no such country exists. Bear with me anyway.

CHAPTER 20. MODULAR ARITHMETIC: TELLING TIME

3 hours after that, though, the clock would reach 0:

$$3 + 4 \equiv 0 \pmod 7.$$

In 6 hours, then, it would be 2 o'clock:

$$3 + 6 \equiv 2 \pmod 7.$$

Likewise, if we started at 3 o'clock and subtracted off 4 hours, we would end up at 6 o'clock:

$$3 - 4 \equiv 6 \pmod 7.$$

We even have the same contrived analogy for multiplication; if we started at 0 and went forward $3 * 10$ hours, we would end up at 2 o'clock:

$$3 * 10 \equiv 2 \pmod 7.$$

There's nothing particularly special about 7 or 12 here; you could do this sort of arithmetic not just mod 7 or 12 but mod any number n that you wanted to use. Although if you're looking for a country where the clocks reset at, say, 2,138,687, you might have to go ahead and found that country on your own.

The important thing to remember here, though, is that working mod n makes all of our arithmetic much easier. As pointed out before, working mod 12 means we only have to worry about the numbers 0-11 in mod 12; similarly, we would only have to worry about the numbers 0-6 in mod 7, or 0-2 in mod 3, or 0-28 in mod 29, or 0-112 in mod 113, or 0-2,138,686 in mod 2,138,687, or...well, you get the point.

20.2 Putting the Mod to Work

In the early 1800's, mathematicians began to realize that this mod business might have other uses besides merely allowing them to determine what time it was and how much longer they'd have to wait until lunch. The first non-temporal advance in modular arithmetic came in 1801 when Carl Friedrich Gauss[4] made the following discovery:

---

[4]We've mentioned Gauss several times already - he was the guy who had a tendency to make major mathematical advances at absurdly young ages. This particular discovery about modular arithmetic appeared in Gauss' 1801 book *Disquisitiones Arithmeticae*, a magnum opus in which Gauss detailed many of the great breakthroughs he had accomplished in his career. He was 24 at the time.

CHAPTER 20. MODULAR ARITHMETIC: TELLING TIME

Gauss' Discovery: *Modular arithmetic is really, really good at identifying equations that have <u>no</u> solutions.*

For example, let's say you are presented with the following equation:
$$x^6 + x^4 - 3x^2 + x^2 = 2.$$

Quick, find an integer solution.
Still looking? Yeah, you'll be doing that for a while.
What if you tried to examine this equation mod 3? It would look like this:
$$x^6 + x^4 - 3x^3 + x^2 \equiv 2 \pmod{3}.$$

As I pointed out at the end of the last section, putting things into Mod 3 World means that we only have to deal with the numbers 0, 1, and 2. Let's plug in 0, 1, and 2 for x on the left-hand side and see what comes out:

$$0^6 + 0^4 - 3(0^3) + 0^2 = 0,$$
$$1^6 + 1^4 - 3(1^3) + 1^2 = 0,$$
$$2^6 + 2^4 - 3(2^3) + 2^2 = 60 \equiv 0 \pmod{3}.$$

None of these are 2 (mod 3). So there are no solutions mod 3.
The reason that this is useful is because Gauss realized the following:

Gauss' Conclusion: *If there are no solutions mod 3 then there are no integer (or even rational) solutions to the original equation.*

This means that we have a really easy way to show that an equation has no solutions: you can show that there are no solutions mod 3, and....actually, there's no "and." You would be all done.

In fact, there's nothing special about mod 3. If you can find that there are no solutions mod 5, or mod 6, or mod 2,381, or mod whatever number you want to deal with then there are no solutions to the original equation. In other words,

Gauss' Broader Conclusion: *Pick your favorite number and call it n. If there are no solutions mod n then there are no integer or rational solutions to the original equation.*

20.3 Not Taking "No" For An Answer

What about cases where there *are* solutions mod n, though? For instance, let's say I gave you the equation

$$y^2 = x^3 + x + 2.$$

If we throw this into mod 3 world

$$y^2 \equiv x^3 + x + 2 \pmod{3},$$

we find that solutions actually exist. Specifically, if you plug in 1 for x and 1 for y, you have

$$1^2 \equiv 1^3 + 1 + 2 \pmod{3},$$

or

$$1 \equiv 4 \pmod{3}.$$

1 is indeed the same thing as 4 mod 3 (if your clock resets at 3, 4 o'clock and 1 o'clock are the same thing). So we have a solution mod 3.

Working in mod 5 land yields a similar phenomenon: if we set up our equation mod 5

$$y^2 \equiv x^3 + x + 2 \pmod{5}.$$

and we plug in 4 for x and 0 for y then we have

$$0^2 \equiv 4^3 + 4 + 2 \pmod{5},$$

or

$$0 \equiv 70 \pmod{5}.$$

0 and 70 are the same mod 5, so we have solutions mod 5 as well.

"Okay," you might be thinking. "It seems like there actually are solutions to this equation mod n. What do we get to conclude from that?" Sadly, the answer is less than satisfactory:

Gauss' Lack of Conclusion: *If there are solutions to the equation mod n then we have no idea what to expect.*

You can see why this might be problematic. Even today, there's still no overarching principle for how one gets around this lack of information; instead, we're forced to pick away at individual cases, and we've only managed to find success in the most rudimentary classes of equations.

CHAPTER 20. MODULAR ARITHMETIC: TELLING TIME

The point of the Birch-Swinnerton-Dyer Conjecture is that it attempts to improve Gauss' Lack of Conclusion for a large class of equations: namely, all elliptic curves. To do this, Misters Birch and Swinnerton-Dyer posed an algorithm that (if correct) would be able to directly translate statements about mod n into statements about the number of rational solutions to a given equation. In the next two chapters, I'll show you where this algorithm comes from, how it might work, and what it would tell us about elliptic curves. Unless clock arithmetic tells me that it's time for lunch. Then all bets are off.

Chapter 21

L-Functions: Convoluted Functions with Weird Powers

21.1 Tick Tock, The Party Don't Stop: Elliptic Curves and Modular Arithmetic

In the last chapter, we talked about what it meant to do things *mod n* and how we might possibly be able to use those ideas to get a sense for how many solutions there are to an equation. We also talked about clocks and how arithmetic works on them, if I recall correctly. That was cool.

In this chapter, we're going to take a look at how one can apply these modular arithmetic ideas to elliptic curves. This will make clocks relevant to the current section[1], which is fun[2].

To do this, let's get back to our favorite lucrative function:

$$y^2 = x^3 + x^2 + x + 6,$$

One thing we could ask about this function would be, "How many solutions does this have mod 2?" The answer is 2; $(x, y) = (0, 0)$ and $(x, y) = (1, 1)$ are both solutions, and it turns out that these are the only ones. We mathematicians express this by saying "N_2", or "the number of answers mod 2", is 2. To make this less wordy, we write

---

[1]Good thing, too, because otherwise I would never have been able to ham-handedly shoehorn a Ke$ha reference in the section title.

[2]By the way, if you're not young enough to get the reference in that last footnote, I guess you could pretend I said "Color Me Badd" instead of "Ke$ha." Maybe that helps?

$$N_2 = 2.$$

We could also ask, "How many solutions does this have mod 3?" or "How many solutions does this have mod 5?" or even "How many solutions does this have mod 7?" (note that we just do primes because they are awesome and tell us everything). If we do so, we get

$$N_3 = 2,$$
$$N_5 = 5,$$
$$N_7 = 2,$$
$$\dots$$

and so on.

Now, if any of these N's were zero, we would know that there are no rational points on this curve (because of the aforementioned Gauss' Conclusion), and we'd be done. However, even if none of them are zero, we still get some nice parting gifts. Specifically, the N's above give a sort of DNA-looking sequence for our elliptic curve:

$$2, 2, 5, 2, \dots$$

Some mathematicians call this the *L-Sequence* for the curve. Other, more verbose mathematicians call it the *Complete List of the Number of Points on This Elliptic Curve Modulo Each Value, Excluding Composite Values for Canonical Reasons, Arranged in Increasing Order of the Prime Value Modulo Which the Number of Points Was Found* (or CLNPTECMEVECVCRAIOPV-MWNPWF, for short). In the interest of keeping this book under 5,000 pages, we'll use the former.

What's interesting here is that the parallels to DNA go a bit further. Specifically, if I said, "I have a curve whose L-sequence is 2, 2, 5, 2,...." (and I listed out the rest of the values), you could say, "Why, that has to be our old friend $y^2 = x^3 + x^2 + x + 6$." In other words, much like snowflakes, fingerprints, and explanations for the JFK assassination, every L-sequence is unique.

Fortunately, this is where the parallels to DNA end. Let's be honest, the analogy is feeling pretty strained already.

21.2 Let's Put These In An Equation and Call It a Day

Now that we've got this list of numbers, we ask, "What's a fun thing to do with a list of numbers?" The answer, as I'm sure you'd agree, is "Put them together into some sort of equation." In fact, that's exactly what we'll do. We'll even give it a name; we'll call it $Z_E(s)$, which means "The Z-function coming from the elliptic curve E. And also, s is involved somehow."

Recall from the Riemann Hypothesis chapter that Euler had introduced the following function:

$$Z(s) = \left(\frac{1}{1-\frac{1}{2^s}}\right)\left(\frac{1}{1-\frac{1}{3^s}}\right)\left(\frac{1}{1-\frac{1}{5^s}}\right)\left(\frac{1}{1-\frac{1}{7^s}}\right)\left(\frac{1}{1-\frac{1}{11^s}}\right)\left(\frac{1}{1-\frac{1}{13^s}}\right)\cdots$$

Well, these $Z_E(s)$ are what this function would look like if Euler, when setting up the above equation, had gotten drunk and spilled ink all over the page:

$$Z_E(s) = \left(\frac{1}{1-(2-(N_2-1))2^{-s}+2^{1-2s}}\right)\left(\frac{1}{1-(3-(N_3-1))3^{-s}+3^{1-2s}}\right)$$
$$\times \left(\frac{1}{1-(5-(N_5-1))5^{-s}+5^{1-2s}}\right)\left(\frac{1}{1-(7-(N_7-1))7^{-s}+7^{1-2s}}\right)$$
$$\times \left(\frac{1}{1-(11-(N_{11}-1))11^{-s}+11^{1-2s}}\right)\left(\frac{1}{1-(13-(N_{13}-1))13^{-s}+13^{1-2s}}\right)$$
$$\times \cdots\cdots$$

It still has a pretty clear pattern; it's just that that pattern is, as the kids say these days, "hella ugly[3]." Note that this series is wholly dependent on which curve you take, since it is determined by the number of solutions that a curve would have mod 2, 3, 5, etc., so, just as in the case of the sequences before, each elliptic curve has its own unique $Z_E(s)$.

Now, this function is certainly visually unpleasant. However, it is irresponsible of us purely to judge a function by its aesthetics without searching for its inner beauty, for sometimes the ugliest functions can lead to the most beautiful results. As such, it is entirely possible that this unsightly function

---

[3]I should note that this definition isn't 100 percent complete, although it's close enough for the purposes of exposition. There are a couple extra factors that are ignored here. I wouldn't worry too much about it if I were you.

CHAPTER 21. L-FUNCTIONS: WEIRD POWERS

is actually an elegant manifestation of some deep and fundamental relation in mathematics.

That's not the case, here, though. This function is terrible.

21.3 Why So Useless? A Confessional

If you asked this function why it is so useless, it would probably break down in tears because, seriously, that's a pretty mean thing to ask. Then, it would probably tell you the following three things:

1.) It's ugly. But you knew that.

2.) It doesn't do anything cool. Remember how when Euler defined his Z-function, he found things like $Z(2) = \frac{\pi^2}{6}$? Well, this one doesn't do anything like that.

3.) You know how we said that $Z(s)$ wasn't defined for any s less than 1? Well, these functions aren't even defined for s less than $\frac{3}{2}$, which is way worse[4].

For a mathematician, that third one is unacceptable. We can deal with a lack of aesthetics, and we can certainly deal with a lack of coolness[5], but if something isn't even defined most of the time, we're bored, and if even the mathematicians think you're boring...well, that's a bit like the Unabomber calling you "too creepy." There's really no down from there.

21.4 Extreme Function Makeover: Elliptic Curve Edition

Now, this is usually the point where some 19th century mathematician like Riemann is compelled to come running in to the story and say, "Wait! This blows up for negative numbers! I have to find a way to patch this function so that it stops blowing up so much!", and then spits out some bizarre magic function that makes everything work. Unfortunately, in this case, no 19th century mathematician came to our rescue. In fact, no 20th century mathematician really came to our rescue, although one mathematician (Helmut

---

[4]This is hopefully the part where you remember that $\frac{3}{2}$ is bigger than 1. Right?
[5]We are mathematicians, after all. Coolness is alien to us.

CHAPTER 21. L-FUNCTIONS: WEIRD POWERS

Hasse, the most famous mathematician ever to be named after a piece of protective headgear) gave pretty good guesses of what these patches should look like, and another (Max Deuring, which sounds like a stage name) proved that Hasse's proposed patches worked in some specific cases. It wasn't until 2001 that the quartet of Christophe Breuil, Brian Conrad, Fred Diamond, and Richard Taylor finally proved that a patch existed for every $Z_E(s)$ by proving the so-called *Modularity Theorem*, a famous theorem that warrants a separate chapter[6] and will be described in more detail if I ever get around to typing up that write-up of Fermat's Last Theorem[7].

Anyway, as is always the case for these sorts of things, we say, "It turns out that there's a patch after all!", and then we write down some incomprehensible miracle function which fixes everything:

$$L_E(s) = \prod_v (1 - q_v^{-s} Frob_v | \lim_{\overleftarrow{l^n}} E[l^n]^{I_v})^{-1}.$$

Yep, there it is.

Much like in the Generalized Riemann Hypothesis section, this new function involves the letter L. Unlike the GRH section, though, this function doesn't involve Dirichlet, so we leave him out of this. As a result, $L_E(s)$ is usually called an *elliptic curve L-function*.

In this case, our new L-function is even more impressive a patch than previous L-functions:

Awesome Fact: *For any elliptic cure E, $L_E(s)$ is defined for **every single** s. Even s = 1. Put that in your pipe and smoke it, Riemann.*

However, it turned out that, having created this $L_E(s)$, the Modularity Fun Boys had gotten a bit more than they bargained for, as they had unleashed upon the world a function that was more powerful than anyone realized. In a turn upon the master usually only seen in kung fu movies and Star Wars, $L_E(s)$ developed a powerful mind control over the elliptic curve E that spawned it. Things were about to get weird....

---

[6]A quick note about this theorem: by the time that Breuil and friends came around, we knew exactly what the patch functions should look like, but we didn't know whether they would always work. *That* was what the Modularity Theorem actually proved here; the patches always work.

[7]Which will not happen in this book.

Chapter 22

BSD: Carrying the One

Having successfully constructed these new L-functions, researchers soon began to realize that each L-function bears an interesting relation with the elliptic curve that spawns it. The key, as it turns out, was to look at what the function did at $s = 1$:

Unbelievably Important Question: *What happens to the function $L_E(s)$ when $s = 1$? Or, more succinctly, what's $L_E(1)$?*

"Why did we pick this value?", you might be asking. To that, I say, remember Riemann's Zeta Function $\zeta(s)$? Remember how that guy wasn't defined when $s = 1$? Well, when mathematicians realized that these L-functions were actually defined at $s = 1$, they said, "There must be something interesting going on at $s = 1$ that makes these functions different from ζ. We shall investigate this. To the grant-proposal writing machine!"

The investigation began, and soon, two mathematicians (Bryan Birch and Peter Swinnerton-Dyer, whose names you may recognize from the chapter title) found that there appears to be a very interesting relation between the L-function at $s = 1$ and the original elliptic curve from which the L-function was obtained. Armed with this new information, Birch and Swinnerton-Dyer put together a grant proposal for the ages; they laid out one of the greatest conjectures of all time, and then, to make sure that the proposal stood out to reviewers, they adorned the cover with My Little Pony stickers. Reviewers were reportedly impressed by the stickers (which were, by all accounts, adorable), but it was the conjecture itself that truly caught the eye of every mathematician in the field:

Chapter 22. BSD: Carrying the One

Peter Swinnerton-Dyer (left) and Byron Birch are two of the greatest clock readers of our time.

Birch-Swinnerton-Dyer Conjecture: *If $L_E(1) = 0$ then the elliptic curve E has infinitely many rational points. If $L_E(1)$ is not zero then the elliptic curve E does not have infinitely many rational points.*

That's pretty amazing. We've been looking for a way to tell if a curve has infinitely many points or not, and it turns out that the answer is probably sitting right there in the L-function we created. In other words, if this conjecture is true then instead of searching endlessly for rational points, we can just compute one value of our L-function and find out everything we need to know.

Unfortunately, we don't know for certain whether this is true; after all, if we did, we'd probably no longer call it "conjecture." However, the Coates-Wiles Theorem, discovered (as you might imagine from the name) by John Coates, Dick Gross, Victor Kolyvagin, Karl Rubin, Andrew Wiles[1], and Don Zagier, proved that we can get at least partway there:

Coates-Wiles' Partial Answer: *If $L_E(1)$ is not zero then the elliptic curve E does not have infinitely many rational points. If $L_E(1) = 0$ then we're not sure.*

So if we're trying to find elliptic curves with infinitely many points, simply evaluating $L_E(1)$ gives us an easy first line of attack where we can separate

---

[1] Remember the Andrew Wiles that I mentioned in the preface to BSD? Same guy.

out many of the curves that won't fit this criteria. It's not perfect, but it's definitely a start.

Chapter 23

Digression: Equations and Diagrams That Are Required to Go in Any Write-up of BSD

I feel obligated to take a little bit of time out of my explanation of BSD to reflect on the absurdity of some of the most important equations involved in this conjecture.

The first equation that I wanted to highlight is one of the primary equations in the statement Birch-Swinnerton-Dyer Conjecture. I include it here mostly because it looks hilariously confusing:

$$\frac{L^{(r)}(E,1)}{r!} = \frac{\#\text{Ш}(E/\mathbb{Q}) \cdot \Omega_E \cdot Reg(E) \cdot \prod_{p|N} c_p}{(\#E(\mathbb{Q})_{\text{Tor}})^2}.$$

I mean, come on, they had to use Roman, Greek, *and* Russian letters? Really? Two alphabets weren't enough?

Incidentally, if that equation isn't long enough for you, there's always this one, which seems to involve every possible permutation of E, n, and K imaginable:

$$\begin{array}{ccccccccc}
0 & \longrightarrow & E(K)/nE(K) & \longrightarrow & \mathrm{Sel}^{(n)}(E/K) & \longrightarrow & \mathrm{III}(E/K)[n] & \longrightarrow & 0 \\
& & \| & & \downarrow & & \downarrow & & \\
0 & \longrightarrow & E(K)/nE(K) & \longrightarrow & \mathrm{H}^1(K, E[n]) & \longrightarrow & \mathrm{H}^1(K, E)[n] & \longrightarrow & 0 \\
& & \downarrow & & \downarrow & & \downarrow & & \\
0 & \longrightarrow & \prod_v E(K_v)/nE(K_v) & \longrightarrow & \prod_v \mathrm{H}^1(K_v, E[n]) & \longrightarrow & \prod_v \mathrm{H}^1(K_v, E)[n] & \longrightarrow & 0
\end{array}$$

I assume that's all clear.

Anyway, let's move on to the second half of the BSD exposition....

BSD II: The Problem Strikes Back

Chapter 24

More BSD: The Stronger, Better, Faster Version[1]

All right, now that we've got the basic idea of what the Birch-Swinnerton-Dyer conjecture is all about, it's time to let these L-functions run loose and see the full strength of what they can do. In this chapter, we'll rev the L-functions up, point them directly at elliptic curves, and see what kind of carnage results[2]. It'll be awesome.

In order to get into the crux of this chapter, you'll remember how in the last chapter we came up with the following correspondence:

BSD Conjecture (i.e. Cliffs Notes for Last Chapter) *If $L_E(s) = 0$ when $s = 1$ then there are infinitely many points on your chosen elliptic curve. If not, there are finitely many points.*

If that had been the end of the story for L-functions and elliptic curves, the BSD Conjecture would have been remarkable enough and likely worthy of a reasonably large cash prize. But Birch and Swinnerton-Dyer wanted to ensure that their conjecture would be worthy of even larger cash prizes, and so they attempted to draw up an even deeper connection between L's and E's. Their improved connection centered on two key observations:

---

[1] I chose this title for the sole purpose of getting that Kanye West song stuck in your head. Harder, Better, Faster, Stronger...

[2] Or something.

1.) Not all infinite sets are alike, so perhaps we could get more information than simply "infinite" or "not,"

2.) There's probably more that you can say about a function than whether or not it's zero.

As they began to discuss these ideas, Birch and Swinnerton-Dyer realized that the resultant information from the two observations was actually very closely related; certain types of infinite sets seem to correspond *exactly* to certain behaviors of the L-function. This relationship seemed mysterious and yet simple and fundamental, which, of course are exactly the same adjectives that are used to describe nearly every important conjecture in number theory (and hence a good sign that the discovery was important - if the same descriptors apply to both your conjecture and, say, the Riemann Hypothesis, you're probably doing something right.) And thus, the stronger, more million dollarish version of the Birch-Swinnerton-Dyer Conjecture was born.

24.1 Questions for Discussion

Now, in order for us to get to the bottom of the two observations above and properly appreciate Birch/Swinnerton-Dyer's realization, we will have to answer two questions:

- "What is the structure of an infinite set?"
- "How zero is zero?"

What both questions have in common is that they sound like the sort of dumb questions that a person would ask after ingesting significant amounts of hallucinogenic substances. That's not really a fair or complete characterization of these questions, though, as it ignores the fact that they're also really opaque and very difficult to answer because they're so ambiguous. One of our goals for this chapter, then, will be to ask these questions in a way that's better for answering them. Another goal would probably be to make them sound less like quotes from a Cheech and Chong movie.

Chapter 24. More BSD!

Dude, what if zero and infinity were like, um, ...wait, what was I saying?

Chapter 25

Elliptic Curve Structure: Like Regular Addition But With Way More Symbols

25.1 Complexity is Complicated!

First, let's take care of this structure of infinite sets silliness.....

Remember that for our various applications of elliptic curve, our goal was to find a curve that had infinitely many rational points so that we had lots of options for setting up whatever real-world applications we had in mind. So let's say that you've found one. Hooray! You can go ahead and set up your computer encryption system, secure in the knowledge that your credit card data and high scores on Minesweeper are now secure from outside attack.

But how safe is your system? Not all infinite sets are alike, after all. If you're trying to set up a cryptosystem that would confuse would-be hackers and you had your choice of infinite sets, you might want to skip this one

$$\{1, 2, 3, 4, 5, 6, 7...\},$$

in favor of one where the pattern is harder to discern, like this one:

$$\{3, 4, 6, 8, 12, 14, 18, 20, 24, 30, 32, 38...\}$$

Conversely, you may want to avoid something that's too hard for even your computer to deal with, like

$$\{5, 7, 38, 104, 12.7, -\pi, 6, x^2 + 1, \mho, \circledast, \frac{\bar{\partial}^2}{7}...\}.$$

CHAPTER 25. ELLIPTIC CURVE STRUCTURE   118

So there's a bit of a balancing act. While it's exciting for us to find an elliptic curve that has an infinitude of points, we want to have some sort of understanding about the structure and complexity of the set.

25.2 Mordell Makes a Mathematical Mend

This problem of depiction was finally solved by Lewis Mordell, a mathematician who came to prominence in the early '20s. Mordell, like most mathematicians of his age, was a huge fan of Minesweeper, and he was anxious to find a way to determine the complexity of an elliptic curve so that he could protect his vital data.

The key insight on Mordell's part came when he decided to start studying the following pairs equations that we've all felt the need to study at some point:

$$X = \left(\frac{3x^2+1}{2y}\right)^2 - 2x,$$

$$Y = y + \left(\frac{3x^2+1}{2y}\right)\left(\left[\left(\frac{3x^2-1}{2y}\right)^2 - 2x\right] - x\right).$$

and

$$X = \left(\frac{y_2-y_1}{x_2-x_1}\right)^2 - x_1 - x_2,$$

$$Y = \left(\frac{y_2-y_1}{x_2-x_1}\right) - \left[x_1 - \left(\left(\frac{y_2-y_1}{x_2-x_1}\right)^2 - x_1 - x_2\right)\right] - y_1.$$

"Oh, that makes perfect sense!" said Mordell because he was much better at mathematics than we are. "That gives us the answer completely!" Then, realizing he would have to explain his ideas to others, he decided to write down what it was that made perfect sense to him.

Let's say you have an elliptic curve. Remember that elliptic curves are things that have y^2 equal to x^3 plus some other x terms. Just so that we can have an example to work with, we'll choose this one:

CHAPTER 25. ELLIPTIC CURVE STRUCTURE

$$y^2 = x^3 + x + 7.$$

The first step is generally to see if we can find *any* rational points on the curve. This isn't as interesting a question as whether we can find infinitely many rational points, but, hey, we have to start somewhere. In our current case, we're in luck because there's a point that's easy to find: if we plug in $x = 1$ and $y = 3$, we find that our equation above works. Thus, we have a point at $(x, y) = (1, 3)$. So far, so good.

Now, here's where the interesting thing happens. Take $x = 1$ and $y = 3$ and plug them into the first pair of weird equations that Mordell was looking at. You end up with

$$X = -\frac{14}{9},$$
$$Y = \frac{35}{27}.$$

You know what that gave us? Another point on our elliptic curve! If you plug in $x = -\frac{14}{9}$ and $y = \frac{35}{27}$ on our elliptic curve, you find that the left equals the right.

Now, guess what happens if you take $(x, y) = (-\frac{14}{9}, \frac{35}{27})$ and put it into Mordell's first pair of equations! You may not want to compute this because it's going to get ugly, but not to worry - I have a calculator handy:

$$X = -\frac{584,761}{44,100},$$
$$Y = \frac{449,103,509}{9,261,000}.$$

Yep - that's another point on the curve! (Trust me on this one.) In fact, you can keep doing this, and you'll keep finding new and different points on the curve.

But wait - there's more! Mordell realized that if you have two different points on the curve (like $(1, 3)$ and $(-\frac{584,761}{44,100}, \frac{449,103,509}{9,261,000})$), you can plug them into the second set of equations[1]:

$$X = -\frac{270,655,811}{552,579,049},$$
$$Y = \frac{32,842,283,080,449}{12,989,475,704,843}.$$

Do you know what this gives us? A mess. Seriously, this thing looks like a dog's meal in mathematical form. But it's also a point on the curve!

In short, Mordell had stumbled upon a sort of elliptic curve addition; he could take a point and "add" it to itself (i.e. plug it into the first pair of equations) to get a new point, or he could take two points and "add" them together (i.e. plug into the other pair of equations) and find another one. He cleverly decided to officially call this process "elliptic curve addition," mostly because he had already devoted all of his creativity to figuring out the actual addition and had very little left for the naming process[2].[3]

25.3 Out of One, Many

All right, so if you fell asleep somewhere in the middle of the last section, here's what you need to know:

Summary for the Last Section: *If you find one rational point on your elliptic curve, you can use it to quickly find many other points on the curve. Also, you should probably go brew yourself a pot of coffee or something.*

If you find one point, you can add it to itself to get more points, and you can add those points to themselves to get still more points.

Mordell said, "Well, that's interesting. I certainly like to be able to find lots of points on the curve!" But then he discovered that there was an im-

---

[1] Here, the first pair is being plugged in as x_1 and y_1, and the second pair is x_2 and y_2. Just thought I'd clarify in case you happen to want to do those computations yourself.

[2] It should also be noted that Mordell was famously inept at naming things. Don't believe me? Just ask his three children, Boy, Girl, and Boy #2.

[3] For what it's worth, the equations that Mordell was considering make a bit more sense when described in terms of geometry instead of algebra. I didn't want to go too much into the geometric explanation here because it takes away from the story, but there's a good explanation of elliptic curve addition with pretty pictures in Appendix C.

CHAPTER 25. ELLIPTIC CURVE STRUCTURE

portant follow-up question:

Mordell's Follow-Up Question: *If you start with a single rational point and do your addition stuff, can you get all of the other points on the curve? Or are there going to be some that you miss?*

and the natural follow-up to the follow-up:

Mordell's Follow-Up Follow-Up Question: *If not, how many points do you need to start with? Can you start with two points? Three? Four?*

Of course, the answer is invariably something like, "It depends on the elliptic curve," which is annoying because it means more work for us. In the case of the curve we chose as our working example for this section ($y^2 = x^3 + x + 7$), one point is actually enough - if you start with the point $(1,3)$, you can get all of the other points on the curve. In other cases, you need two points. Sometimes even three. Someone actually recently found a curve where you need more than 27 points (we don't know the exact number of points needed for that one, we just know that it's more than 27)[4]. So it'll obviously vary a bit from curve to curve.

25.4 Intuition Through Things That You Know: Even Numbers

All of this stuff with elliptic curves and structure and points and elliptic curves is probably new to most readers of this chapter, so let's develop some intuition by thinking about something you've seen before: the (positive) even numbers.

Hopefully, you envision the positive even numbers to be something like this:

$$\{2, 4, 6, 8, 10, ...\}$$

There are a couple of things to notice about this set. if we start with 2,

---

[4]This one is actually the world record right now. It was found by Noam Elkies, a name that you may have seen because I mentioned him in the introduction as well. He's really good at computing these sorts of things.

we can get every other number in the set by just adding 2 to itself a bunch of times. Actually, wait. That's the only important thing to notice. So go ahead and notice that.

In other words, if I wanted to get the number 56 (which is obviously one of the numbers in the set), I could just take $2 + 2 + 2$.... a whole bunch of times and eventually end up with 56. Or, I could add $2 + 2$ to get 4, and then $4 + 4$ to get 8, and $8 + 8$ to get 16, and $16 + 16$ to get 32, and then $32 + 16$ to get 48, and finally $48 + 8$ to get 56. The same would be true of any number in this set; if you said, "Give me the following number!" and then named a number in the set, I could just start adding 2 to itself a whole bunch of times until I either got to the number or got bored and told you to do it yourself, or I could strategically keep adding numbers that I've gotten from previous iterations of adding things until I end up in the right place.

So if in the spirit of Mordell, I asked, "How many numbers do we need to start with in order to get all of the even numbers?," you would say, "We only need one number: the number 2. If we start with 2, we can get the rest of the set by adding 2 to itself repeatedly (or by cleverly iterative addition starting with 2)." Sometimes, to describe this phenomenon, we say that 2 *generates* the even integers.

25.5 Negatives Always Make Things Harder

On the other hand, what if I took my set above and included negatives? And zero? You know, like

$$\{...-10, -8, -6, -4, -2, 0, 2, 4, 6, 8, 10, ...\}$$

Does adding 2 to itself a bunch of times still generate this set? No! Of course not. That would be silly. No matter how many times you add 2, you'll never end up with, say, -12. So just starting with 2 isn't enough to give us the rest of the set.

Well, what else could we add in to get the rest of the set? -2 would seem to be a good option. (In fact, -2 is the right answer.) If we start with 2 and -2 we quickly see that anything else in the set can be written as either a bunch of 2's added together or a bunch of -2's added together. (Or, in the case of 0, one of each, since $2 - 2 = 0$.) In this case, we need two numbers at the beginning (2 and -2) in order to generate the rest of the set[5].

---

[5]Mathematical aside: when I say, "the number of points/numbers/elements you need,"

CHAPTER 25. ELLIPTIC CURVE STRUCTURE

This is the game that Mordell was trying to play here[6]. Sometimes, like in the case of the positive even numbers, you just need to start with one point on an elliptic curve to generate all the others. (This happens fairly often.) Sometimes, you need two or three. Or four. Or maybe even 5. But no matter what elliptic curve we start with, there's always some finite list of points that we can start with that will generate the rest of the points on the curve.

25.6 Ready for Rank!

So excited were mathematicians by Mordell's discovery that they rushed out and named it:

Definition: *Let's say you've picked your favorite elliptic curve and named it E. Additionally, let's say that E has infinitely many rational points, which is always exciting. In that case, the number of points you need to start with to generate the all of the rational points on the curve is called the* **rank** *of the curve E.*

I'm not entirely sure why we mathematicians chose the word "rank" here. Maybe the people who named it were fond of the miliary? I dunno.

Anyway, rank is actually a really good description of the complexity of an elliptic curve. If I asked you, "How complicated is your elliptic curve E?", you might say, "Probably pretty complicated, because it took you about seven chapters to describe how to describe complexity." But if you got past that, you might also say something like, "Well, today I chose a curve that's pretty complicated! It has rank 10." And I would say, "Yep. That's pretty complicated." So that's exciting - we can communicate to each other in math now.

---

what I really mean is "the minimum number of points/numbers/elements you need *assuming you've chosen wisely*." I could instead have picked 4, 0, and -2 to generate all the other numbers in this set, but that wouldn't have been particularly smart of me.

[6]Well, that and Minesweeper

Chapter 26

How Zero is Zero?

Now that we've dealt with this structure business, we finally get to explain just how zero a zero can be.

To do this, we'll introduce a couple of equations that will serve as useful examples. We'll name them $f(x)$, $g(x)$, and $h(x)$ after three famous mathematicians, Fred, Greg, and Hector[1]:

$$f(x) = x - 1,$$
$$g(x) = (x-1)(x+2)(x-3),$$
$$h(x) = (x-1)^2(x-4)$$

To begin, let's just look at the first equation, $f(x)$. Notice that if I plug in 1 for x, this equation becomes 0. To express this fact, mathematicians would say that $f(x)$ *has a zero at* $x = 1$.

Of course, this isn't unique to the first equation. In fact, all of these equations have this property in common; if I were to plug in 1 for x, all three equations would evaluate to 0. In other words, $f(x)$, $g(x)$, and $h(x)$ all have a zero at $x = 1$. You might notice that $g(x)$ and $h(x)$ have zeroes at other points as well, but don't be concerning yourself with that, OK? One thing at a time here.

Once we've established that these equations all have a zero at $x = 1$, there's an obvious question,

---

[1] Specifically, Fred Euler, Greg Euclid, and Hector Einstein.

CHAPTER 26. HOW ZERO IS ZERO?

Obvious Question: *Which of these equations has the best zero?*

However, we'd prefer not to answer this question because it's really judgmental, and we like to be accepting of all numbers and see the beauty that each one brings. Except 5. 5 can go suck an egg.

Let's change the question to something a little less judgmental:

Less Mean But Still Obvious Question: *Which of these equations has the strongest zero?*

That's better.

To measure this, watch what happens when we divide all three of our expressions by $x - 1$ and let cancelation do its thing:

$$f(x) : \frac{\cancel{x-1}}{\cancel{x-1}} = 1,$$

$$g(x) : \frac{\cancel{(x-1)}(x+2)(x-3)}{\cancel{x-1}} = (x+2)(x-3),$$

$$h(x) : \frac{(x-1)^{\cancel{2}}(x-4)}{\cancel{x-1}} = (x-1)(x-4)$$

Now, take a look at what we have left. What's left of $f(x)$ and $g(x)$ (1 and $(x+2)(x-3)$, respectively) no longer have a zero at $x = 1$, as their zeroes were unable to weather the attack by $x - 1$. Hector's function $h(x)$, by contrast, did better; what remains of $h(x)$ (which is $(x-1)(x-4)$) still evaluates to zero when you plug in $x = 1$, meaning that Hector survived the attack with its zero at $x = 1$ intact. It stands to reason, then, that $h(x)$, the only function whose zero didn't perish in the great battle of $x - 1$, had the stronger zero.

Thus, we have our answer:

Answer: *While all of the zeroes are important in their own way and should be appreciated for their individual virtues and capabilities, $h(x)$ had the strongest zero, bearing in mind that strongest is merely a statement of physical strength and should not be taken as a value judgement meant to disparage the other two zeroes or discourage them from being proud of who they are.*

CHAPTER 26. HOW ZERO IS ZERO?

26.1 Measuring Zeroes: A New Cult Arises

As it turns out, we actually have a measurement for how much division by $x-1$ an expression can survive before its zero meets an untimely demise. We call this measurement the *order of the zero*. The Order of the Zero sounds like it should be some mysterious monastic club with bizarre entry requirements and secretive meetings by candlelight, but really it's just a measuring stick[2]. Too bad, huh?

Actually, wait a minute. If I declared the Order of the Zero to be an actual monastic order, I could get donations and followers and tax exemptions and stuff, right? In that case, ignore the statement above. The Order of the Zero is an extremely mysterious body of which very little is known to outsiders. To find out how to unlock these mysteries, please send a large donation to the author of this book.

Anyway, let's write down exactly how this measurement works:

Definition: *Say that we have an expression that evaluates to zero when we plug in 1 for x. The **order of the zero at** $x = 1$ is the number of times you can divide the expression by $x - 1$ before the expression no longer has a zero at $x = 1$.*

So if we looked at the expression $(x-1)^4(x+2)$, we would say that it has a zero of order 4 at $x = 1$, since we could divide by $x - 1$ four times before we finally got something other than zero when we plugged in $x = 1$.

Addendum to Definition: *If the function doesn't have a zero at $x = 1$, we sometimes say that it has a zero of order 0, since the number of times we had to divide by $x - 1$ was 0.*

Of course, there's nothing special about $x = 1$ here; we could have talked about zeroes at $x = 2$ (which would require us to divide by $x - 2$ until the zero went away) or any other number instead.

I should note something of importance here: not every equation in the world is a nice, easy expression like $(x-3)^2(x+2)^5$ where we can figure out

---

[2]Interesting and completely true fact I learned while researching this book: Unlike the Order of the Zero, which is not an actual order, The Order of the Trapezoid is actually the governing body for the Church of Satan. I don't know about you, but I am pleased to learn that my long-standing suspicion of geometers was entirely justified.

the order of the zero at $x = 3$ by just looking at the exponents and saying, "Hey, the exponent attached to $(x - 3)$ is 2. I bet the order of the zero at $x = 3$ is 2. And...we're done." Sometimes, the order is a little better hidden in the expression, which means that we actually have to think or do a little math before we actually find our answer. That's fine, though; the method is still the same, and if you're averse to "doing a little math," well, you probably don't want to choose mathematics as your vocation in the first place.

Note also that we didn't have to use the letter x; we could have used any letter. Like t. Or y. Or even....s. And with this, we segue back into talking about L-functions...

Chapter 27

Finally: The Full, All-Powerful, Earth-Shattering, Cavity-Reducing, Baldness-Curing Birch-Swinnterton-Dyer Conjecture

Now that we have these definitions, we can finally state the full Birch-Swinnerton-Dyer Conjecture.

The idea is this: remember how the L-function $L_E(s)$ would sometimes evaluate to zero when $s = 1$? Well, fortunately, we just spent several pages explaining what we can do with that. If we check out the order of the zero at $s = 1$, something really cool happens:

The Full, All Powerful Birch-Swinnerton-Dyer Conjecture: *Remember what the rank of an elliptic curve is? Good.*

Now, remember what the order of a zero is? No, not the religion, the other one - you know, the number that tells us how strong a zero is.

If you have an elliptic curve E and you went and made an L-function ($L_E(s)$) from it then the following absolutely shocking equation holds every

CHAPTER 27. THE FULL BSD

single time:

$$[rank\ of\ E] = [the\ order\ of\ the\ zero\ of\ L_E(s)\ at\ s = 1].$$

To put this more simply, the higher the order of the zero, the more complicated the elliptic curve.

In other words, not only does evaluating the L-function at $s = 1$ tell you whether the elliptic curve has infinitely many points or not (which is already useful), it actually tells you exactly the structure of the infinite set you're talking about by telling you exactly what the rank is. Basically, this says that the L-function tells you just about everything you could ever know about elliptic curves - assuming the conjecture is right, of course.

To put it bluntly, that's nuts. On the one hand, you have this made-up version of addition and you're looking at this curve's structure and how points add together. On the other hand, you have this function (which wasn't even defined half the time until we fixed it) that evaluates to zero at some random point. And yet...they give exactly the same information.

27.1 How Close Are We?

Now that we've got the whole statement worked out, you might be wondering, "How close are we to getting this conjecture? It's a nice conjecture and all, but if we can't prove it, it doesn't do us a ton of good."

Well, we're getting there. I mentioned a result by Wiles and Coates (and a bunch of other mathematicians) after stating the weak conjecture; now that I've gone and stated the full conjecture, I can actually tell you the result in a little more detail:

Coates-Wiles' Partial Answer (the REAL Story): *Here's what we know, or at least what Coates and Wiles figured out in 1977*:

- If the order of the zero for $L_E(s)$ is 0 then the rank of E is also zero.
- If the order of the zero for $L_E(s)$ is 1 then the rank of E is also 1.
- If the order of the zero for $L_E(s)$ is something bigger than 1 then we're stuck.

This may not seem all that impressive until you consider the fact that in

1977, while Coates and Wiles were discovering this major theorem, many of their friends and colleagues were hitting the Cheech and Chong pretty hard, if you know what I mean.

Wait, that wasn't what I meant to say at all. Ignore that last statement.

What I meant to say was that the Coates-Wiles Theorem may not seem all that impressive except that mathematicians have come to suspect that the vast majority of L-functions have a zero of order either 0 or 1, so this theorem is believed to cover most of the elliptic curves.

If I'd written this chapter a couple of years ago, I would have stopped here, but then something big happened in 2010. After almost 35 years of Coates and Wiles having the best BSD-related result, a Princeton professor named Manjul Bhargava and his erstwhile graduate student, Arul Shankar, dethroned Coates-Wiles by proving the following:

Bhargava-Shankar Theorem of Getting A Quarter of the Way There: *The BSD Conjecture holds for at least 25 percent of all elliptic curves.*

Of course, people always say, "The first 25 percent is always the easiest[1]," but the theorem above is probably a good indication that we're on the right track; in fact, recent results have used their methods to push the number up to about 66%. This theorem still doesn't give any indication as to what happens with the other 34 percent or so (in fact, there are limitations to how far this Bhargava-Shankar attack could be pushed - we know that getting 100 percent would be out of the realm of possibility for their methods), but given that the previous record was 0 percent, we're all pretty excited about the progress.

---

[1] Here, when we say "people", we mean "no one."

Chapter 27. The Full BSD

Not John Coates and Andrew Wiles.

Chapter 28

Appendix C: An Actual Explanation of Elliptic Curve Addition

In an earlier footnote, I claimed that the method of addition that Mordell discovered is much easier to understand if I start drawing graphs than it is if I actually try to explain it with words and algebra. Well, it's time to back up those words with action, so in this appendix, I'm going to draw pretty graphs and make you understand how elliptic curve addition is done. I should note that "adding points on a curve" sounds like a deep and intimidating idea, but, in reality, all we're doing is finding a way to combine two points to get another point; if we can do that, we can apply our new definitions of generator and rank and form cults and make money and stuff.

First, in order for this appendix to make any sense, we're going to need a visual of what elliptic curves look like, so let's draw those out. Basically, there are two options; usually, an elliptic curve either looks like the nose and mouth of a sideways frowny face:

CHAPTER 28. APPENDIX C: ELLIPTIC CURVE ADDITION 133

Actual Elliptic Curve

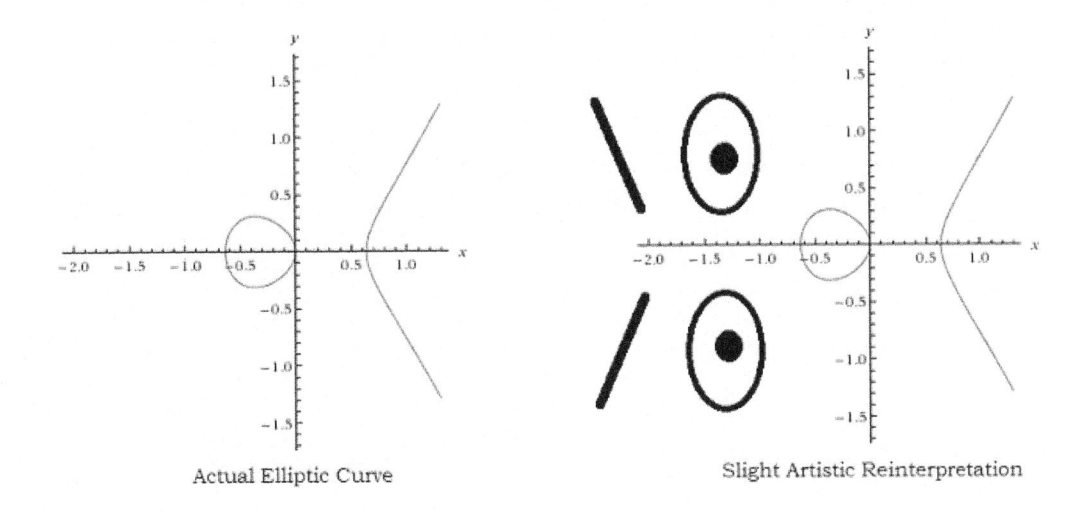

Slight Artistic Reinterpretation

or else it looks sort of like a ghost that has fallen over onto its side for some reason:

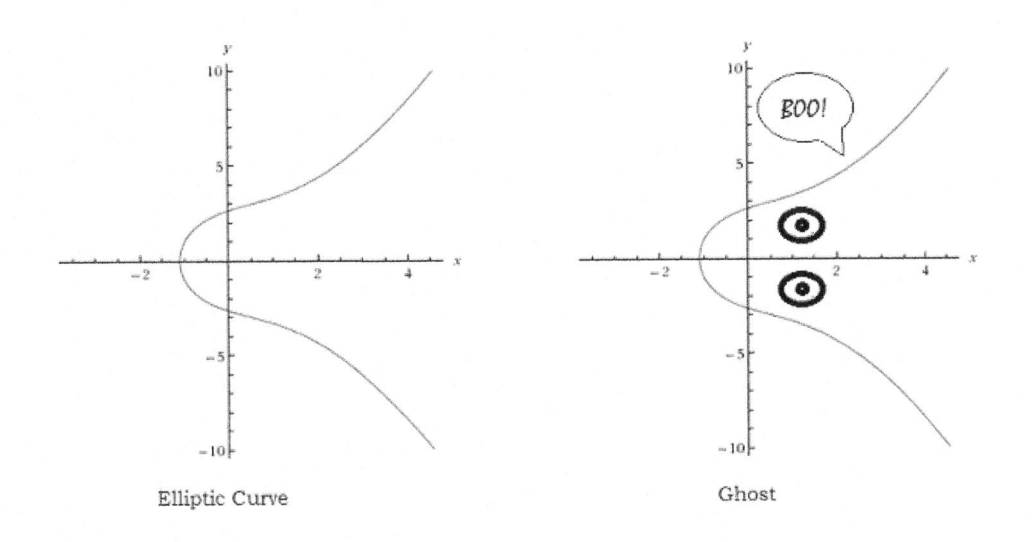

Elliptic Curve

Ghost

CHAPTER 28. APPENDIX C: ELLIPTIC CURVE ADDITION 134

In order to demonstrate our new addition law, let's pick a specific elliptic curve. Actually, let's use the one we talked about in Chapter 19:

$$y^2 = x^3 + x^2 + x + 6?$$

This one turns out to be one of the "fallen ghost" ones, so if I put this into my calculator and say, "Draw!", it comes out like this:

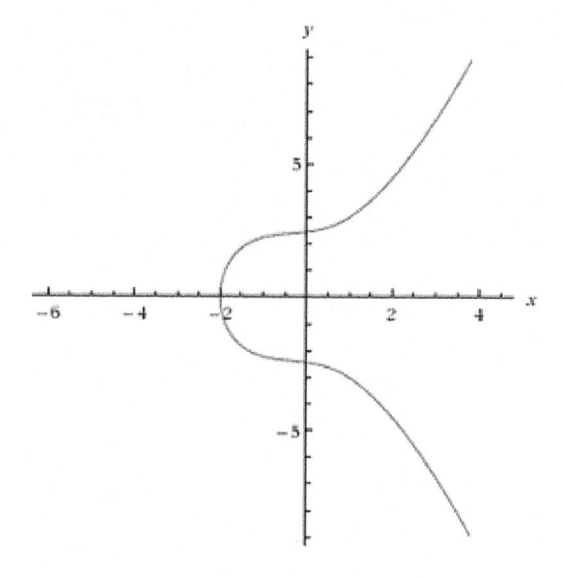

Now, let's say I had two rational points on the curve. I'll call them P for "Point" and Q for, um, "QDifferent Point." Let's plot them:

CHAPTER 28. APPENDIX C: ELLIPTIC CURVE ADDITION

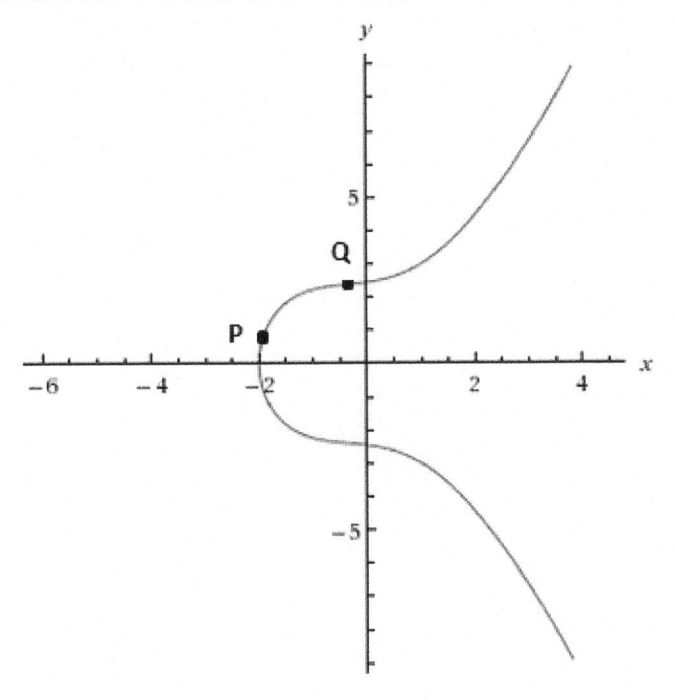

Of course, the whole point of Mordell's work was that I had a way to add them together. The process for doing this is threefold:

28.1 Connect the Dots. La la la la.

The first thing we do is draw a line to connect the dots:

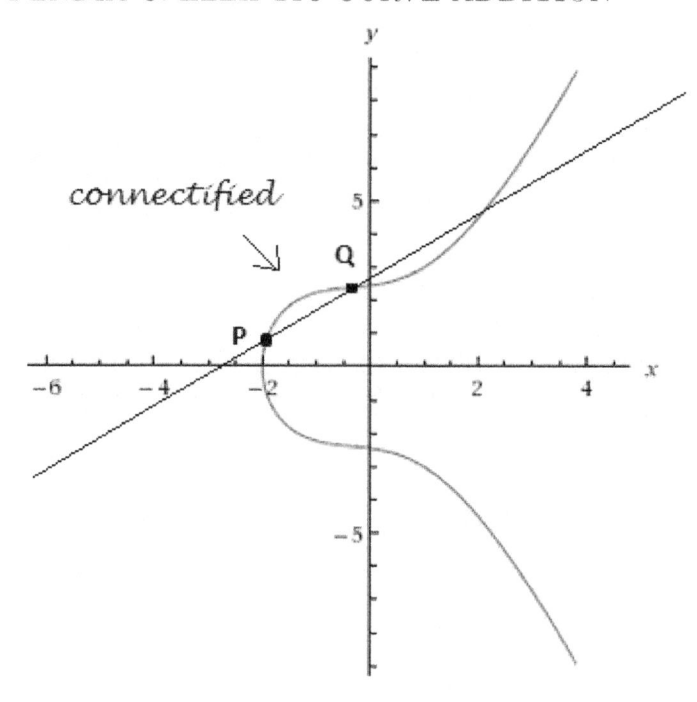

CHAPTER 28. APPENDIX C: ELLIPTIC CURVE ADDITION 136

28.2 Look for the mysterious "Third Point"

Note that when we draw our line, there are three places where the line intersects the curve: the point P, the qdifferent point Q, and one other place. Let's call this third point R in honor of the fact that "R!" is what a pirate would usually say when connecting the points on an elliptic curve.

CHAPTER 28. APPENDIX C: ELLIPTIC CURVE ADDITION

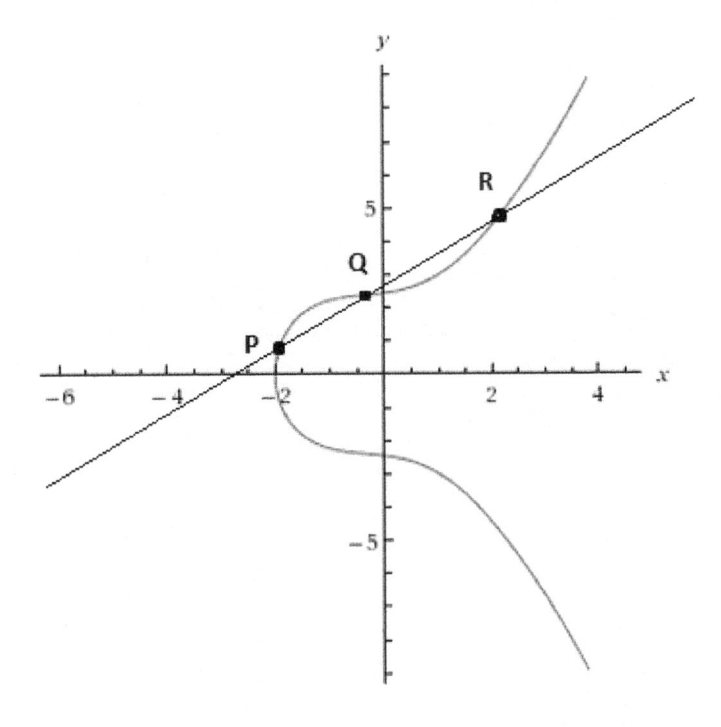

28.3 Drop It Like It's Hot

Now, take the point R and drop it straight down (or, if it's already on the bottom, move it straight up):

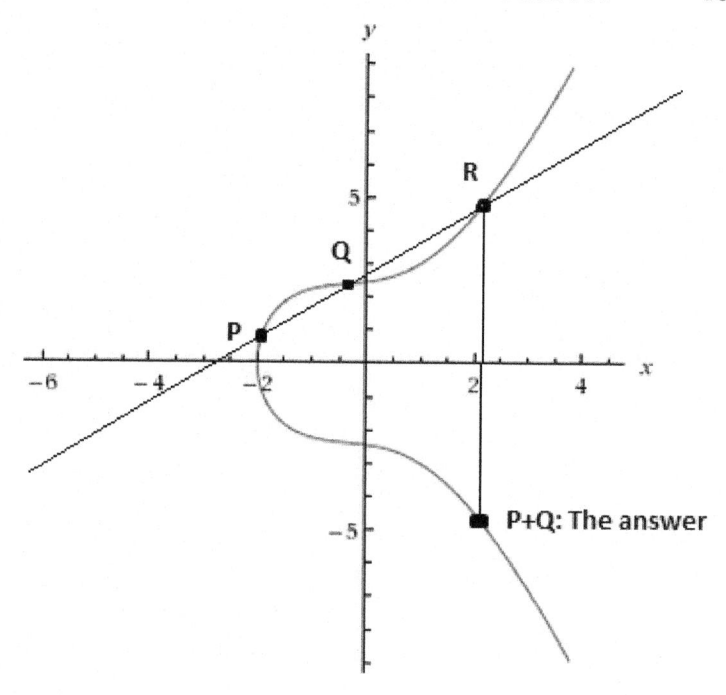

CHAPTER 28. APPENDIX C: ELLIPTIC CURVE ADDITION 138

BAM! That's your answer. It's an easy three-step arts-and-crafts approach to elliptic curves, and yet, somehow, it gives us a method of addition that has all sorts of nice properties. You can try this with any elliptic curve: pick two points on your curve, connect them with a line, follow the line to find the third point, drop that third point, and there you are. If you want, you can even draw some eyes to complete the ghost or frowny face picture when you're done.

28.4 One Small Problem

Of course, there is an issue with this definition: what if you wanted to add a point to itself? Like, what if you wanted to take P and add it to itself (giving you $P + P$ somehow) and forget about Q altogether?

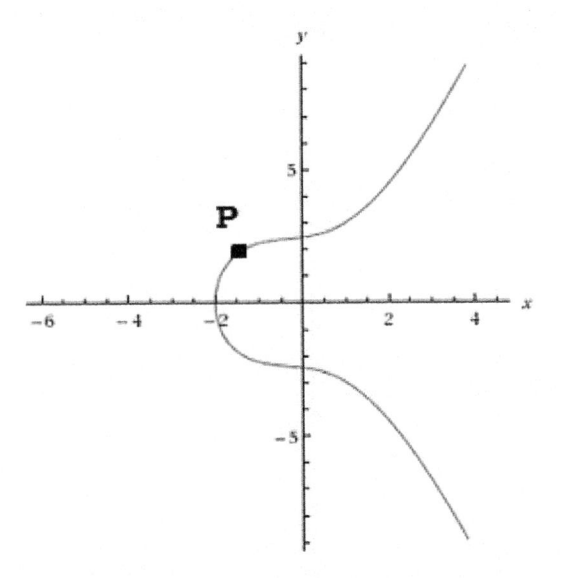

Q is gone, but it is also forgotten.

Well, the obvious problem here is that when you get to the part that says, "Draw a line to connect the dots," you can draw pretty much any line you want through P, since connecting P to P isn't going to be very hard. To combat this, we say, "Draw a line through P that just touches (but doesn't go through) the elliptic curve":

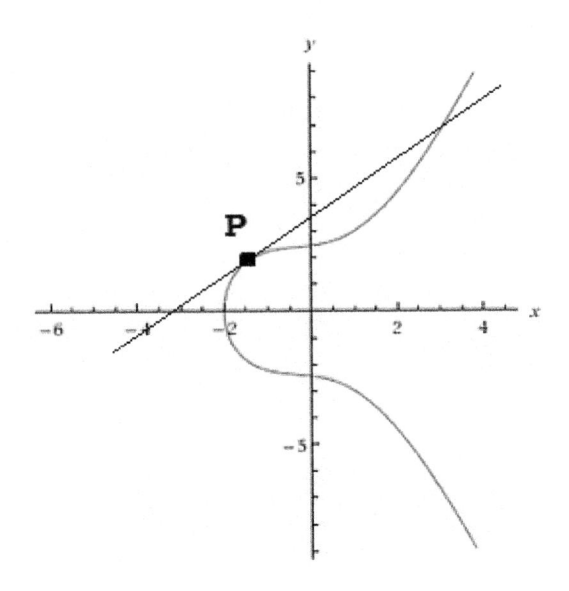

CHAPTER 28. APPENDIX C: ELLIPTIC CURVE ADDITION

Now, we do the same process as before. We find the other point where the line and curve intersect and call it R:

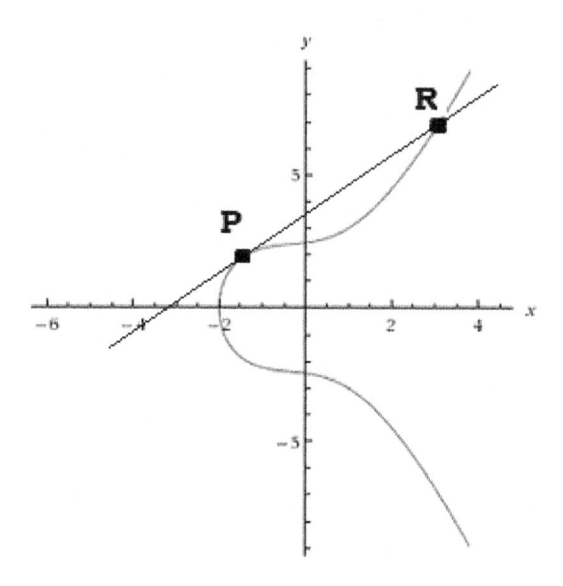

and then we drop R like a bad habit:

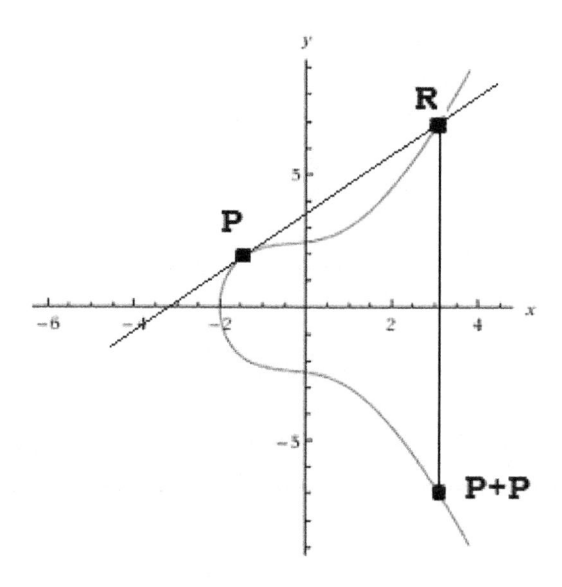

and we've got our answer.

The Erdős Conjecture on
Arithmetic Progressions

Chapter 29

Introducing Paul Erdős

I decided that at some point in this book, we have to talk about Paul Erdős, the brilliant and extremely eccentric Hungarian mathematician who was the forefather of a number of the ideas in this book.

Erdős, who lived from 1913 until 1996, was one of the more unusual people ever to grace mathematics. Stories about his personal attributes, which are entirely true, read like a list of exaggerated stereotypes about mathematicians: he had little interest in anything outside of mathematics, didn't know how to tie his shoes, learned how to butter his own toast at age 23, traveled with his mom everywhere he went, had no sexual interest in either gender, gave away nearly every penny he earned to charities, and took amphetamines so that he could put in 19-hour days of mathematics.

Erdős also happened to be the most prolific mathematician of the 20th century. He worked with hundreds of mathematicians, wandering Europe, America, Asia, and Australia for forty years and going from mathematician's house to mathematician's house solving problems and writing mathematical papers. One of Paul's favorite sayings was, "Another roof, another proof," and it was a pretty apt characterization of much of his professional life. At the time of his demise, Erdős had authored or co-authored (mostly the latter) over 1,500 papers, and, much like Tupac Shakur, papers co-authored by Erdős continued to appear for several years after the mathematician's death.

In fact, Paul Erdős worked with so many mathematicians that a construct known as the "Erdős number" arose to describe how far (mathematically) each mathematician is from Erdős. If you wrote a paper with Erdős, your number is 1; if you wrote a paper with someone who wrote a paper with Erdős, your number is 2; if you wrote a paper with someone with Erdős

CHAPTER 29. INTRODUCING PAUL ERDŐS

number of 2, your Erdős number is 3, and so on. At the time of Erdős' death in 1996, pretty much anyone who was anyone in the math world had an Erdős number of 4 or less. In fact, from personal experience, I can say that as a mathematician, it is pretty much a guarantee that any non-mathematician who knows anything of Paul Erdős will ask me what my Erdős number is, as though I would just happen to know that information off-hand[1].

In other words, Paul Erdős was the prototype for the crazy mathematician. In fact, since he effectively took a vow of chastity *and* poverty, we could pretty much call Paul Erdős the patron saint of the crazy mathematician.

One of the many ideas that Erdős contributed to math was the idea of putting a bounty on problems that he felt should be solved. He picked problems that he wanted solved and then affixed dollar amounts, ranging from $1 (for the easier - though still difficult - ones) to a couple thousand (for the harder ones), that would be given as rewards for the solution to each of the problems. The prizes weren't great, but they were the mathematical equivalent of putting bulls-eyes on certain problems - they let other mathematicians know which problems Erdős thought were a.) hard (but not impossibly so) and b.) important. Incidentally, Erdős never actually bothered to keep track of his finances and, of course, had a tendency to give most of his money away, so it was entirely likely that another mathematician would end up paying the reward that Erdős promised, but that's beside the point. A few of the problems still remain unsolved, and even though Erdős died in 1996, his colleagues have promised that they will honor the awards promised by Erdős[2].

One of the most expensive of these problems is known as the Erdős Conjecture on Arithmetic Progressions, for which the Hungarian mathematician promised $3,000 of somebody's money. Other mathematicians, realizing the difficulty and importance of this problem, have since increased the bounty to $5,000, which is where it stands today. It's not nearly as impressive an amount as the million dollars that the Clay Math Institute promised for the solutions to their problems, but it comes with the Paul Erdős Seal of Approval, which is impressive enough in its own right[3].

---

[1]Five.

[2]Which, I suppose, is no different from when he was alive.

[3]On the topic of Erdős....if you ever get a chance, you should definitely read Paul Hoffman's *The Man Who Loved Only Numbers*, the definitive biography about Paul Erdős. Hoffman traveled around with Erdős for a while and basically just asked all of Erdős' friends for stories about him, which, if you know anything about Erdős, would be sufficient

CHAPTER 29. INTRODUCING PAUL ERDŐS

Paul Erdős: Bounty Hunter

---

fodder for several books. The author also gets into a little bit of the math that Erdős was doing, and he presents it in a pretty accessible way (sort of like this book, but with far fewer movie references and stupid jokes). It's a fascinating read.

Chapter 30

Flip It Good: Reciprocals and Sums

30.1 Walk Like a Reciprocal

Have you ever looked at a bunch of numbers and thought, "I wonder what it would look like if I put those on the bottom of a fraction!"? Of course you have. Don't lie. You do this all the time.

Well, you're not the first one to have had this inclination. In fact, the idea of flipping things upside down goes back to the days of the Egyptians, who were absolutely enthralled by fractions where the top was 1 (like $\frac{1}{7}$ or $\frac{1}{34}$ or $\frac{1}{2}$). In fact, they were so enthused by such fractions that if you gave them a fraction that *didn't* have 1 on the top, they'd split it and contort it until it did; for instance, if you gave them $\frac{31}{83}$, they'd rewrite it as something like

$$\frac{31}{83} = \frac{1}{3} + \frac{1}{25} + \frac{1}{6225}$$

Don't ask me why they did this. I'm guessing they were just compulsive. In fact, I bet they were the sorts of people who, when they emptied the change out of their pockets and put it on the dresser, would sort their change by denomination, then by year, then by cleanliness[1].

Incidentally, this whole splitting of fractions into things with 1 on top is known as *Egyptian fractions*. As you can imagine, this bizarre exercise

---

[1] Now that I think about it, this might explain why they built all of those pyramids - they probably saw a whole bunch of large stones lying around and just felt the need to stack them.

CHAPTER 30. FLIP IT: RECIPROCALS AND SUMS 148

is now considered to have very important uses in such burgeoning modern-day vocations as papyrus-making, corpse-mummifying, standing sideways for portraits, and, of course, pleasing the god Osiris.

30.2 Adding Things

Now, while the Egyptians were all about the flipped number with the 1 on top, it wasn't until the 18th century that people realized that the most exciting thing to do with these things was to add lots and lots of them together. Just like with every other discovery in the 18th century, it was actually Euler who figured this out; while he was playing with his new $Z(s)$ function that we discussed in the Riemann Hypothesis chapter, he came to the realization that addition was where the party was at[2].

It all started when Euler tried to add the following[3]:

$$\frac{1}{1} + \frac{1}{2} + \frac{1}{3} + \frac{1}{4} + \frac{1}{5} + \frac{1}{6}....$$

Surprisingly, this went off to infinity. I should clarify - this isn't surprising to you *now* because you already read the Riemann chapter where I told you that this was the case. However, Euler, for some reason, hadn't yet read my book, so he was forced to figure this out for himself.

Why is this so surprising? Well, the terms individually get smaller and smaller, i.e. each term in this sequence is closer to zero than the one before it. However, the whole sum still blows up; the terms don't get small fast enough to counteract the fact that you're adding infinitely many of them.

What if we didn't use all of these fractions? For instance, what if we only used the fractions that had prime numbers in the denominator:

$$\frac{1}{2} + \frac{1}{3} + \frac{1}{5} + \frac{1}{7} + \frac{1}{11} + ...$$

It turns out that this goes off to infinity as well. It actually does so really, really, *really* slowly[4], but much like the inspirational children's story of The Little Engine That Could, this series keeps chugging along, climbing higher

---

[2]Somewhere, my high school English teacher's hand just hit her forehead.

[3]You might remember this - it's $Z(1)$ from Euler's Z-function.

[4]If you're wondering how slowly this thing blows up, let me put it this way: if you added up the first trillion trillion fractions in this series, the sum would be about 4. I mean, this thing really takes its sweet time getting to infinity.

CHAPTER 30. FLIP IT: RECIPROCALS AND SUMS 149

and higher, ignoring the mean other fractions that laugh at him and tell him that he'll never make it, persevering until, finally, the Engine blows up from overwork and goes off to the great "infinity" in the sky. Actually, wait. That might not have been how the story went. Anyway, this sum blows up, too.

In fact, even if we just restricted ourselves to the primes that have a 4 as one of their digits:

$$\frac{1}{41} + \frac{1}{43} + \frac{1}{47} + \frac{1}{149} + \frac{1}{241} + \frac{1}{347} + \frac{1}{349} + \frac{1}{401} + \ldots$$

we would *still* be left with something that blows up.

Do all such sums go off to infinity? Well, it depends what you mean by "all such sums," but it is certainly possible to come up with sums that look sort of like the ones above but don't actually go to infinity. For example, let's say we used only denominators that are perfect squares (like 1, 4, 9, and 16, which are 1^2, 2^2, 3^2, and 4^2, respectively). Then our sum would look like this:

$$\frac{1}{1} + \frac{1}{4} + \frac{1}{9} + \frac{1}{16} + \frac{1}{25} + \frac{1}{36}\ldots$$

This one doesn't go to infinity at all; in fact, it goes to $\frac{\pi^2}{6}$. In this case, you're still adding infinitely many things together, but the individual terms grow smaller and smaller quickly enough to avoid the blowup fate that befell the train engine.

The same sort of thing would happen if we only took fourth powers (i.e. 1^4, 2^4, 3^4, etc.):

$$\frac{1}{1} + \frac{1}{16} + \frac{1}{81} + \frac{1}{256} + \frac{1}{625}\ldots$$

This guy goes to $\frac{\pi^4}{90}$, which is most decidedly not infinite.

As such, we have our exciting conclusion: sometimes, these sums blow up, and sometimes they don't. Deep, huh?

30.3 Ooh - Crash and Burn

OK, I'll admit it: the ending to that last section was pretty pathetic. I'm not sure what happened there. My apologies. In penance, I'll try to restate what

CHAPTER 30. FLIP IT: RECIPROCALS AND SUMS 150

we did above in a way that makes it way less boring and far more worthy of ancient Egyptian worship.

To begin, remember that our first example of one of those infinite sum thingys was

$$\frac{1}{1} + \frac{1}{2} + \frac{1}{3} + \frac{1}{4} + \frac{1}{5} + \frac{1}{6}....$$

Like a powder keg with a hair-triggered time bomb on pins and needles just waiting for the other shoe to spark, it eventually blew up. Or, in words that involve fewer mixed metaphors and more mathematics and alliteration,

Mercifully Mixed Metaphor-Free Mathematical Formulation of Foregoing Fraction Phenomenon: *If you take the set*

$$\{1, 2, 3, 4, 5, ...\},$$

put each term over 1, and add everything together, you get something that blows up.

The point of writing our observation in this way is that in discussing whether these sums go off to infinity or not, we can probably stop talking about the 1's on top and just list the denominators, like

$$\{1, 2, 3, 4, 5, ...\},$$

or

$$\{1, 4, 9, 16, 25, ...\}.$$

The 1's aren't changing or going anywhere, so let's admit to ourselves that it's only our choice of denominator that matters and move on.

In fact, let's move on to the point where we assign some nomenclature to these sets:

New, Natural Nomenclature for Noted Numerical Notion: *If we have a set of denominators (like the one above) where taking 1 over each term and adding everything together gives a blowup, we will call the set a* **dynamite set**. *If we have a set where this activity does not create a sum*

CHAPTER 30. FLIP IT: RECIPROCALS AND SUMS

*that blows up, we will call the set **dud set**[5].*

So the set
$$\{1, 2, 3, 4, 5, ...\}$$
falls under our new "dynamite" definition because the sum
$$\frac{1}{1} + \frac{1}{2} + \frac{1}{3} + \frac{1}{4} + \frac{1}{5} + \frac{1}{6} ....$$
blows up. Similarly, the sets
$$\{2, 3, 5, 7, 11, 13....\}$$
and
$$\{41, 43, 47, 149, 241, ...\}$$
can also be called dynamite because they're explosive as well. However, the sets
$$\{1, 4, 9, 16, 25, 36....\}$$
and
$$\{1, 16, 81, 256, 625....\}$$
are both duds.

For the rest of the chapter, we're mostly going to ignore the duds because, honestly, duds sets are boring[6]. Instead, we'll focus on the dynamite side of things. In particular, we're going to sort through the dynamite and see if we can spot any cool patterns or consistent characteristics that come with having an explosive set.

---

[5] You might be surprised to learn that this isn't standard terminology. This is in large part because there isn't any standard terminology, and I've actually heard a couple of different proposals for what to call these sets. The most common verbiage I've heard is to call a dynamite set a "large set" and a dud set a "small set." Small and large? Booorrrring!!!

More recently, I heard a good suggestion from Pete Clark, a mathematician at the University of Georgia, who suggested that dynamite sets be called "substantial" (and dud sets would presumably be "not substantial"). However, I'm going with my verbiage because, honestly, it's more fun to say "dynamite."

[6] I mean, I called them "duds." That should tell you that they're not that interesting.

Chapter 31

Arithmetic Progressions: The Godwin's Law of Mathematics[1]

31.1 Primes: Are They Clumpy?

Have you ever watched one of those History Channel shows where they take every third letter of every fourth page of the Bible and then read it backwards, and somehow it vaguely foretells the rise of Hitler if you ignore a few of the random letters that don't make any sense? Well, that's the sort of the idea behind arithmetic progressions.

I'll explain what I mean. Invariably, whenever someone looks at a list of primes (or whatever set they're dealing with), they start looking for patterns. For some reason, that's what the human brain does - it tries to make order of whatever it sees, even if what it sees has no order to speak of. Since we're dealing with the primes, the person will find something like this:

$$5, 11, 17, 23, 29$$

and say, "Hey, check this out! There's a pattern! If start at 5 and then keep

---

[1] About the title here: if you've never heard of Godwin's Law then today is a big day because you're about to learn one of the fundamental laws of the universe. Godwin's Law (in the paraphrased form that I'm stealing from Wikipedia) states that "given enough time, in any online discussion - regardless of topic or scope - someone inevitably criticizes some point made in the discussion by comparing it to beliefs held by Hitler and the Nazis."

As a corollary, if you are arguing a topic with someone and the other person makes an analogy to Hitler or the Nazis, you are allowed to declare that you have won the argument by Godwin's Law.

CHAPTER 31. ARITHMETIC PROGRESSIONS: GODWIN'S LAW

adding 6, you keep getting primes! Just look - you hit 11, then 17, then 23, then 29. Maybe we can keep adding 6's forever and find primes!"

Of course, the punchline is that 35, the next number in the sequence, isn't a prime, but it's too late - your friend has already made a fool of himself. Ha ha! Sucks to be him.

Apart from embarrassing himself, though, your friend has stumbled on an interesting phenomenon - a clump consisting of five equally spaced primes (which is kind of neat-looking). So that's fun.

Now, what if, instead of five, we wanted to find larger clumps of equally spaced primes? For instance, is it possible to find six equally spaced primes? Yes it is:

$$7, 37, 67, 97, 127, 157$$

This sequence stops at 157 (since 187 is not prime), but we still hit our target - we've got a list of six primes, and each one is 30 more than the one before it.

What about seven equally spaced primes? Sure - why not:

$$7, 157, 307, 457, 607, 757, 907$$

Eight? You bet:

$$199, 409, 619, 829, 1039, 1249, 1459, 1669$$

You might be wondering whether we could keep finding larger and larger clumps of equally spaced primes. Well, by now, you know the deal; before we start attacking math problems, we have to give everything a name:

Definition: *Let's imagine we have a set that has, say, 7 equally spaced numbers. We would then say that our set has an <u>arithmetic progression of length 7</u>.*[2]

Now, just to make things look cooler and mathier, let's replace the 7 with n. So, we'll say we had n equally spaced numbers in our set. These n numbers would then be called an <u>arithmetic progression of length n</u>.

So our sequence of

$$5, 11, 17, 23, 29$$

---

[2] I should note that in this context, we actually pronounce "arithmetic" as "a-rith-MET-ic," not "a-RITH-me-tic" like you learned in grade school.

would be an arithmetic (or arithMETic) sequence of length 5, while the one with

$$199, 409, 619, 829, 1039, 1249, 1459, 1669$$

is an arithmetic sequence of length 8.

OK. *Now*, we can go ahead and ask the question we've all been waiting for. Our set here is the prime numbers, and we'll be looking to see if we can find arithmetic progressions in the primes.

Question: *Pick a big ol' n. Like 481. Or 58,3325. Or even 337,693,127. Make it as big as you like.*

Is there some way to guarantee that no matter how big an n I choose, I will always be able to find an arithmetic progression of length n? And would this foretell Hitler's rise to power?

As it turns out, the last question isn't that interesting, given that we already know pretty much about Hitler's rise to power and such a prophecy wouldn't be of much use anymore[3]. We'll concentrate on the rest of the question instead.

The non-Hitler part of this question was posed by Erdős in the mid-1930's. It turned out to be a tough one, but it finally fell in 2004 when two mathematicians, Ben Green and Terence Tao, opened up a can of Proof on this baby:

Green-Tao Theorem: *Pick a big number. No, a bigger one. Make it so big that you don't even feel like writing it down, so you'll just call it n for short.*

No matter what n you've chosen, you can still find an arithmetic progression of length n amongst the prime numbers. In other words, no matter how big a clump of equally-spaced primes you want to find, you'll always be able to do so.

Booyeah! We in the math community were pretty excited when this result came out. In fact, it's one of the major reasons that Terence Tao won

---

[3]Honestly, I'm not quite sure why those History Channel people are so concerned about prophesizing Hitler's rise to power. Wouldn't it be more useful to prophesize something that *hasn't* already happened? We've pretty much got all of the info on Hitler by now.

CHAPTER 31. ARITHMETIC PROGRESSIONS: GODWIN'S LAW

the Fields' Medal (the math version of the Nobel Prize, except they only award it once every four years[4]) in 2006.

In other words, if we wanted to find an arithmetic sequence of length 21, Green and Tao say that we can do it, and with a little work, we find that they're right:

$$28112131522731197609,$$
$$28112131522731197609 + 9699690,$$
$$28112131522731197609 + 9699690 \cdot 2,$$
$$28112131522731197609 + 9699690 \cdot 3,$$
$$....$$
$$28112131522731197609 + 9699690 \cdot 20$$

In fact, if I wanted to find an arithmetic sequence of length 5,000,000,000,000, I could do that, too. It just so happens that I have no wish to do so. If you want to find one, though, go right ahead. Just be warned: it'll probably take a while to write down.

---

[4]In case you're not aware, there's no Nobel Prize in mathematics. Know why? Neither does anyone else.

Chapter 31. Arithmetic Progressions: Godwin's Law

Terence Tao unleashes yet another proof on the world.

Chapter 32

Back To The Dynamite

32.1 Just How Special Are The Primes, Really?

OK. So we've got this thing that's cool about the prime numbers; specifically, you can find huge clumps of equally spaced primes (as huge as you like, in fact). Tao and Green proved it, and they knew what they were talking about, so we're all set there. However, this actually brings up an interesting question: is there something unique and special and possibly even mystical about the primes that makes them have large clumps, or is this just something that would happen if we have any large set?

Actually, let's be a little more specific. Remember that the set of primes

$$\{2, 3, 5, 7, 11, 13, 17, 19, 23, 29, 31, 37, 43..\}$$

were a dynamite set - which is to say, if you added

$$\frac{1}{2} + \frac{1}{3} + \frac{1}{5} + \frac{1}{7} + \frac{1}{11} + ...,$$

you'd find your sum going off to infinity. Well, it's not the only dynamite set in the world; there's also the one we wrote earlier:

$$\{41, 43, 47, 149, 241, ...\},$$

as well as many, many, many others.

Now, let's say we had a dynamite set. What we'd like to know is if we can find large clumps of equally spaced numbers in each of these sets as well, i.e. the following:

Question About Arithmetic Progressions: *If you have a set that is dynamite, is that set guaranteed to have an arithmetic progression of length 7? What about 8? 9? 10? Are you guaranteed to have an arithmetic progression of <u>any length you choose</u>?*

Answer: *Yes. Or possibly no. One or the other. Either way, it's worth $5,000 to whoever can prove it.*

It's not a million dollar problem, but most of us wouldn't complain about being handed $5,000 to do something that we're already trying to do anyway.

32.2 Why We Care

This actually leads to an interesting meta-question of sorts. As was mentioned before, whenever we humans face some sort of large pool of data, such as the primes or the arrangement of letters in the Bible, we instinctively look for patterns, and, if we find something that looks like a pattern, we immediately assume that there's some sort of underlying structure to whatever we're looking at[1]. Of course, sometimes we're right - that's pretty much how we discovered things like physics and biology. On the other hand, sometimes, we're just making stuff up - for example, take a look sometime at the *spooky* parallels that people have discovered between the Kennedy and Lincoln assassinations[2] and you'll realize that there are a lot of people with a keen interest in history and way, way, way too much time on their hands.

Moreover, if you've got a large pool of data, you're almost guaranteed to have something that looks like a pattern. For instance, if you looked at all of the stars in the sky, you're almost guaranteed to have some (say, eight or so) that fall in a straight line just by random chance. Pointing to these eight stars that fall in a line and saying, "See? Someone put those in order!

---

[1]This is actually such a common occurrence for humans that we've coined a word for this phenomenon; the act of ascribing patterns to randomness is called "apophenia."

[2]Both of their last names were seven letters long! They both had vice-Presidents named Johnson! Both were shot in the head! On a Friday! One was shot in a theater and the perpetrator was caught in a warehouse, while the other was shot from a warehouse and the perpetrator caught at a theater!
Coincidence? You bet.

CHAPTER 32. BACK TO THE DYNAMITE

Maybe those stars signify an interstellar trade route!" misses the point - an arrangement like this is pretty much guaranteed to happen if you have a lot of dots.

As mentioned before, that's the sort of determination that we're trying to make here. We have this cool thing about primes (they have large, equally spaced clumps). The question, then, is this: is this property something special that indicates some sort of structure to the prime numbers, or is it something that's guaranteed to happen any time you have a large (er, dynamite) set (sort of like seeing eight stars in a line)? It's a good question that would tell us a lot both about the primes and about the sorts of things we can expect for large sets in general.

The other question, of course, is when the History Channel is going to get around to making a special about patterns in prime numbers. It's way more mysterious than the Lincoln/Kennedy stuff or the Nostradamus specials they keep pumping out, and I would *totally* watch.

Problems that are Easy to Understand and Absolutely Impossible to Solve

Chapter 33

Easy to Understand, Impossible to Solve

33.1 Introduction

No discussion of the most important problems in number theory is complete without a description of the simple ones that everyone loves, so here goes....

As the great and slightly insane mathematician Paul Erdős once noted,

"Babies can ask questions about primes which grown men cannot answer. Especially if those babies have Ph.D.'s in mathematics[1]."

As far as I can tell, this is patently false, as most babies I have talked to don't really understand the concept of prime numbers and are mostly interested in teething and soiling themselves, but the point still remains that number theory possesses a number of unsolved problems that are extremely easy to state and understand. In this chapter, we discuss several of the most beautiful and renowned of these conjectures.

Unfortunately, these simple problems are of somewhat limited interest to mathematicians as far as actually attempting to find solutions because the problems are extremely hard and no one has any idea how to attack them. In fact, some of these problems have been around for over 2,000 years and *still* no one has any idea how to do them. These are the sorts of problems

---

[1] I should point out that Erdős actually did say the first of these two sentences. I'm not entirely certain about the second one.

that you absolutely do not put in a research proposal because the proposal will come back with laugh-spittle all over it.

This is not to say that these problems are of no use, however. For one thing, they are very effective in luring talented (and unsuspecting) high school and college students into number theory. A bright young student will start out by looking at a simple proposition about primes or integers that seems like it should be easy to answer (like Goldbach's conjecture or Collatz Conjecture, both discussed in the upcoming pages), and they'll say, "Wow, this number theory stuff is fun and accessible!", and then we number theorists come in and say, "Hey there, little mathematician! Do you want to know the secret to that problem you're working on? Well, it's right here in my office. Why don't you step inside?", and then the student comes into the office and we give them a little shove and, KA-THUNK!, the student goes tumbling down the rabbit hole, and suddenly he or she is thinking about Galois actions in unramified extensions of totally real number fields and wondering what the hell happened to that simple question about primes. It's a ruthless system, but I'll be damned if it isn't effective[2].

Wait, who is the target audience for this book? Talented high school and college students? Um, never mind then. You should probably ignore that last paragraph. Number theory is fun and easy!

Anyway, from a mathematician's perspective, the primary use of these problems is that they give us research direction. It is often the case that someone will say, "I can't prove this whole conjecture, but I can try to prove part of it, or I can prove it in some other case that might give evidence that this conjecture is true." Usually, doing this leads to the development of new tools that can then be used to attack other problems in mathematics, which can be a pretty big deal in its own right; for instance, a partial result toward Goldbach's Conjecture (explained in subsequent pages) significantly advanced our knowledge in an area known as Sieve Theory (not explained in subsequent pages), which has turned out to be really useful for proving all sorts of things about prime numbers.

Regardless, I should remind you that these problems are extremely hard. They've been pondered by millions of mathematicians, some of them absolutely brilliant (others presumably less so), and they're still far from being

---

[2]For the record, my last two high school science fair projects were entitled "Collatz Conjecture" and "Collatz Conjecture II," and here I am. I'm telling you, this system works.

resolved. On the plus side, it is considered a rite of passage for every aspiring number theorist to, at some point in his or her development, posit an incorrect proof of either the Twin Prime Conjecture, the Goldbach Conjecture, or the $n^2 + 1$ problem[3],[4]. So there's that.

All right, enough talking - let's see some problems....

---

[3] Mine was the twin prime conjecture. My proof was brilliant! It was also completely wrong.

[4] Fermat's Last Theorem used to be on this list, but Andrew Wiles decided to be a killjoy and posit a correct proof of this one in 1994. Seriously, way to spoil the fun, Andrew.

Chapter 34

Collatz Conjecture: 1930's Version of Angry Birds

34.1 Lothar Collatz: The Least Interesting Man In the World

Back in the 1920's and 30's, the world was populated by savages who hadn't yet discovered the massive societal value of devoting hundreds of hours to noble endeavors like Angry Birds or Addiction Solitaire. To waste time in those days, you either had to find someone who told interminable stories or find a simple mathematical problem that was as addicting as it was impossible. It was a meager and difficult existence, and, as you'd expect, people soon grew so bored and restless that they decided to go ahead and have a World War just so they'd have something to do.

It was in this world that a young man named Lothar Collatz came of age. Early in life, Collatz realized that his dream was to come up with new and novel ways to waste people's time, and he spent much of his existence dedicated to this task. Although he grew to love the sound of his own voice and eventually learned to talk for days on end about a whole range of painfully uninteresting topics, Collatz soon discovered that if he wanted to find a time waster that would waste time all over the world, his best chance would be via the mathematical route (since it was difficult for him to talk to the whole world at once, particularly before the invention of the television). After a few marginally successful attempts in graduate school, Collatz's masterstroke came in 1937 when he discovered the following algorithm:

CHAPTER 34. COLLATZ CONJECTURE: 1930'S ANGRY BIRDS

34.2 Collatz's Algorithm for Wasting Time

Let's start with a (positive) integer. We'll call it x to make it sound more mysterious.

Step 1: If your number is even, divide it by two (i.e. take $x/2$). If it is odd, take $3x + 1$. This gives you a new number.
Step 2: Whatever number you got from the previous step, call it x. Repeat the previous step - if your new x is even, divide it by two, and if it's odd, take $3x + 1$. This gives you a new number.
Step 2: Whatever number you got from the previous step, call it x. Repeat the previous step - if your new x is even, divide it by two, and if it's odd, take $3x + 1$. This gives you a new number.
Step 2: Whatever number you got from the previous step, call it x. Repeat the previous step - if your new x is even, divide it by two, and if it's odd, take $3x + 1$. This gives you a new number. I assume you're getting the point here.
And so on....

Let's try an example. Er, I mean,

34.3 Let's Try An Example

That's better.
We'll start with 5, which means that $x = 5$. Now, let's try some of these steps:

Step 1: 5 is definitely odd. Do you know what that means? That's right - 5 gets $3x + 1$'ed.
$$3(5) + 1 = 16.$$
Step 2: OK, so far so good. Now, 16 is even, so we divide it by 2, giving us 8.
Step 2: 8 is also even, so we chop it in half, giving us 4.
Step 2: 4 is even. Split that puppy in half; now, we've got 2.
Step 2: 2 is even. Cut that one in half and we get 1.
Step 2: 1 is odd. Do the $3x + 1$ thingy to it:

CHAPTER 34. COLLATZ CONJECTURE: 1930'S ANGRY BIRDS

$$3(1) + 1 = 4$$

But we've already seen what happens with 4; 4 will go to 2, which goes to 1, which goes back to 4, and the cycle continues. So we're stuck in a loop.

What if we started with another number like 7? Well, in that case, the steps give us 22, then 11, then 34, 17, 52, 26, 13 , 40, 20, 10, 5, 16, 8, 4, 2, 1, 4, 2, 1, 4,... and there we are back in the loop again.

What about a bigger number like 15? Starting with 15, we get 15, 46, 23, 70, 35, 106, 53, 160, 80, 40, 20, 10, 5, 16, 8, 4, 2, 1, 4, 2, 1, 4... and we're stuck in that loop once more.

Wait, you might be saying to yourself. Will we always get stuck in this loop? In other words, is it the case that regardless of which number we start with, we will eventually end up with 4, 2, 1,...?

Good question. I certainly have no idea. Fortunately, other mathematicians have also considered this problem. The bad news is that they don't know, either. The good news is that they wasted lots of time working on this problem and getting nowhere - time that might have otherwise been used for the betterment of humanity. Looks like we dodged a bullet there.

In fact, as the great Japanese mathematician Shizuo Kakutani once said,

"For about a month everybody at Yale worked on it, with no result. A similar phenomenon happened when I mentioned it at the University of Chicago. A joke was made that this problem was part of a conspiracy to slow down mathematical research in the U.S."

Or, as colleague Atle Selberg put it more succinctly:

"Go to hell, Collatz[1]."

As such, we are left only with Collatz's tantalizing guess:

Collatz's Guess (or Collatz's Conjecture, if you will): *No matter which x you start with, if you do this algorithm enough times, you will eventually end up with 1, 4, 2, 1, 4, 2, 1,...*

---

[1] Several readers have questioned the veracity of this quote, which makes sense when you consider that I may have made it up.

So it seems like the answer is probably yes, but, really, we're just making stuff up at this point.

The good news, however, is that Collatz did indeed achieve his dream of creating a useless activity that would waste people's time for years to come. Even today, Collatz's algorithm continues to provide a low-tech time-waster for countless mathematicians. Stranded in a boring talk? Pick a number and use Collatz's algorithm and see if it takes you down to 1. Waiting for a ride? Stuck in rush hour traffic? Marooned on a deserted island? Bored at the dinner table? Bust out a pencil and paper and let Collatz take you away. It's good for all occasions.

34.4 Why Do We Care About the Collatz Conjecture?

Because Angry Birds is taking too long to load.

34.5 No, Seriously. Why do we care? Other than the whole timewaster thing, of course

Well, the reason that the Collatz algorithm is so interesting is because it acts so bizarrely and unpredictably. In particular, if I give you a number and say, "Do the Collatz algorithm to it," you have no good way of predicting how long it will take to get stuck in the 4, 2, 1, 4, 2, 1 loop unless you're starting with something dumb like, say, 4. There seems to be no rhyme or reason to it; numbers that are close together like 91 and 93 nevertheless go through vastly different routes to get down to 1 (93 takes 17 steps to get down to 1, while 91 takes a whopping 92 steps and meanders through numerous three and even four digit numbers along the way).

The great mathematician Willie Stargell probably put it best when he noted that trying to guess what Collatz's algorithm would do

> "..is like trying to throw a butterfly with hiccups across the street into your neighbor's mailbox."

CHAPTER 34. COLLATZ CONJECTURE: 1930'S ANGRY BIRDS

I suppose that there are a few people out there who wouldn't really consider Willie Stargell a "mathematician" as much as they would "an outfielder for the Pittsburgh Pirates in the 1970s." Some of these people may even contend that when Stargell made the above quote, he wasn't really talking about the Collatz Conjecture but instead explaining the difficulty of throwing a certain type of pitch. These people are liars. Stargell was talking about the Collatz Conjecture. Period.

Anyway, the fact that the algorithm behaves unpredictably stands in stark contrast to what we know about it, which is that it is an easy algorithm governed by two simple rules. This phenomenon (an algorithm with simple rules nevertheless doing complicated and unpredictable things) is not a rare one, as there are many algorithms that come from all over mathematics that act in similar ways; however, it's one where we're still on the proverbial ground floor in the Tower of Understanding, stuck in the Lobby of Vague Comprehension near the Front Desk of Unincorporated Facts and the Fake Plant of Misunderstood Theorems, ringing the Bell of Trial and Error on the Desk of the Receptionist of New Mathematical Ideas in hopes that he or she will point us to the to the Elevator of Actually Getting Something Accomplished. It makes for an interesting phenomenon as well as a horrible analogy.

In fact, if you've ever studied physics or read Jurassic Park[2], you may have heard of something called "Chaos Theory." When mathematicians/physicians write books for the general public, they tend to "explain" Chaos Theory with impenetrable statements like, "A hummingbird flaps its wings in China, and it causes a hurricane in the US," which sounds more like a fake Confucius quote than an actual scientific statement. What chaos theory actually says is, "If you take a system with just a couple of basic rules and let it run, small changes in the starting position may cause large and unpredictably weird changes in the behavior of the system." The Collatz Conjecture is the number theory version of this phenomenon, and while we have a rudimentary understanding of chaos theory in some contexts, we have almost no feel for how these sorts of things work when we place them in a number theory setting. If we could figure out the Collatz Conjecture, it would be the first time that we number theorists would be able to definitively say *anything* about these sorts of phenomena without resorting to unhelpful similes involving

---

[2]The book, not the movie. The movie pretty much skips the pseudo-science and goes straight for the CGI effects, a decision that I wholeheartedly support.

Willie Stargell brought patience, home run power, and veteran presence to the math world.

butterflies with respiratory problems. And, really, those butterflies have suffered enough.

Chapter 35

Goldbach Conjecture: Everything Breaks into Primes, but in a Weird Way

One of the many surprising ways in which mathematicians resemble other human beings is that in mathematics, just as in other disciplines and walks of life, we occasionally have major breakthroughs and discoveries from people who could be characterized (in colloquial vernacular) as "one-hit wonders." Indeed, while many of the mathematical ideas and conjectures we've discussed in this book came from great mathematicians like Euler and Riemann, mathematics is not without its own versions of Vanilla Ice and The Baha Men - men who, for one shining moment, discovered a universal truth and forced us to re-examine our views of mathematics[1] and then realized how hard it would be to come up with another one of these new insights and thus gave up and went back to playing MarioKart.

Perhaps no man in mathematics history is a more famous example of this phenomenon than Christian Goldbach. Goldbach, in a single inspired moment, managed to formulate one of the two or three most famous conjectures in all of mathematics - and that's pretty much it. In fairness to him, of course, coming up with a conjecture with that level of popularity is an immense accomplishment, so it's not really much of an indictment to say that the conjecture was by far his greatest achievement, but he'll be spending the rest of eternity living off of that thing like Tommy Tutone off of Jenny's

---

[1] Or at least reconsider who it was that let the dogs out.

phone number[2].

This is not to say that Goldbach didn't try again. Goldbach also went on to author a theorem about sums of primes, now known as the Goldbach-Euler Theorem, that is known to tens of mathematicians around the world. Today, this theorem even has the rare distinction of having a Wikipedia page, meaning that it is one of the 100,000,000,000,000,000 or so most important theorems in all of number theory. His official biography also notes that he "studied complex analysis for a while" and "considered planetary motion that one time when he was at the gym", and also that he "discovered more than 15 new uses for twine."

In this chapter, however, we'll limit ourselves to the Goldbach Conjecture and save the Goldbach-Euler Theorem for another, less interesting book. As the story goes, Goldbach was playing with numbers one day[3] when he noticed an interesting pattern involving the even numbers. First, he saw that

$$2 + 2 = 4.$$

On its own, this isn't a particularly surprising result. So he kept going:

$$3 + 3 = 6$$
$$3 + 5 = 8$$
$$5 + 5 = 10$$

"Wait a minute," he said. "It looks like every even number can be written as a prime plus another prime!"

He tried a few more, just to check his hypothesis:

$$5 + 7 = 12$$
$$3 + 11 = 14$$
$$3 + 13 = 16$$
$$7 + 11 = 18$$
$$3 + 17 = 20$$
$$11 + 11 = 22$$

---

[2] Too obscure of a reference? Well, I had to include it - this is a book about math, after all, and there haven't been too many hit songs about numbers.

[3] When playing with numbers, parental supervision is always recommended.

Chapter 35. Goldbach Conjecture: Breakin' 2

Eis, eis, kleine.

"Looks convincing to me," said Goldbach. "I shall make it a conjecture!" And conjecture he did:

Goldbach Conjecture: *Any even number bigger than 2 can be written as the sum of two primes.*

This conjecture, if true, fundamentally alters our understanding of prime numbers. You see, the reason we care so much about primes is because they are the building blocks for multiplication; after all, you can write any whole number as a product of a bunch of prime numbers (often referred to as prime factorization, or, if you want to make math sound more like biology, prime decomposition). What this conjecture says is that primes may also be the building blocks for addition, since every even number can apparently be written as a sum of two primes. In short, this tells us that prime numbers are even more important and fundamental than we thought, which, given the general esteem in which we number theorists hold prime numbers in the first place, is pretty amazing.

35.1 The Ternary Goldbach Conjecture: Weak Sauce

Now, the original Goldbach Conjecture only deals with even numbers, so it occurred to Goldbach that he could stretch his fifteen minutes of fame by making another variant of this conjecture that addressed odd numbers instead of evens. This "odd-number conjecture" came to be known as the Weak Goldbach Conjecture because it looks like Goldbach's Conjecture and also because it's pretty weak. Sometimes, the conjecture is also called the Ternary Goldbach Conjecture because it's pretty ternary as well.

So, what is it? Well, when Goldbach was playing with even numbers and writing them as sums of two primes, he noticed that a similar pattern emerged with the odd numbers (or at least those odd numbers that are bigger than 5):

$$2 + 2 + 3 = 7$$
$$2 + 2 + 5 = 9$$
$$3 + 5 + 3 = 11$$
$$5 + 3 + 5 = 13$$
$$3 + 5 + 7 = 15$$
$$2 + 2 + 13 = 17$$
$$5 + 7 + 7 = 19$$

"Wait a minute," said Goldbach in a moment of deja vu. "Every odd number can be written as a sum of three primes!"

Not bothering to wait for further verification this time, Goldbach plowed ahead and made a conjecture about this as well:

Goldbach's Wussier (er, Weak) Conjecture: *Every odd number that is at least 7 can be written as the sum of three primes.*

You may have noticed that we act rather dismissively toward this conjecture, making fun of its strength and even occasionally placing "Kick me!" signs on its back when it isn't looking. Why, you may be wondering, are we so cruel towards this conjecture?

Well, let's say someone proved the regular-strength Goldbach Conjecture. Hooray! We throw a parade in their honor, erect a statue, name our firstborn after them - you know, the usual hoopla that follows the announcement of a major mathematical result. Now, we can write out the even numbers as

CHAPTER 35. GOLDBACH CONJECTURE: BREAKIN' 2

sums of two primes:

$$2 + 2 = 4$$
$$3 + 3 = 6$$
$$3 + 5 = 8$$
$$5 + 5 = 10$$
$$5 + 7 = 12$$
$$3 + 11 = 14$$
$$3 + 13 = 16$$
$$7 + 11 = 18$$

and so on.

Now, let's make a slight alteration to everything I just wrote above. See if you can catch what I did - I'll try not to make it too obvious:

$$2 + 2\underline{+3} = 7$$
$$3 + 3\underline{+3} = 9$$
$$3 + 5\underline{+3} = 11$$
$$5 + 5\underline{+3} = 13$$
$$5 + 7\underline{+3} = 15$$
$$3 + 11\underline{+3} = 17$$
$$3 + 13\underline{+3} = 19$$
$$7 + 11\underline{+3} = 21$$

Did you catch the trick? It was pretty subtle, but hopefully you still got it.

Anyway, that's the thing about weak Goldbach. Since any odd number can be described as "an even number plus 3," it stands to reason that if we can prove the usual Goldbach Conjecture about even numbers, we can also prove the weaker one by just adding 3 to every even number. It's just that easy[4]!

Now, you might be asking, "Why do we bother with the weak Goldbach Conjecture if it just follows easily from the regular one?" The reason is quite

---

[4]It should be noted that this implication doesn't go in the other direction; knowing something about the weak Goldbach doesn't give any information about the strong Goldbach conjecture.

simple: no one has the slightest idea how to prove the regular Goldbach Conjecture, so saying that "the weak Goldbach Conjecture is easy if you can prove the regular one!" is kind of like saying, "Making money on the stock market is really easy if you can build a time machine!" or "Basketball is really easy if you can figure out how to shoot jumpshots from half-court!" If we're looking to prove things about prime numbers adding up to other numbers, weak Goldbach is a far better place to start; it's far more approachable because it affords so much more flexibility, since you have three primes to play around with instead of two.

35.2 Is It True?

A funny thing happened recently with the ternary Goldbach Conjecture. After years of being "the version of the Goldbach Conjecture that seemed doable," the weak Goldbach Conjecture underwent a facelift and became "the version of the Goldbach Conjecture that's been proven." In other words,

New Theorem: *The weak Goldbach Conjecture is actually true. Every odd number is the sum of three primes.*

Pretty amazing, no? This all happened pretty recently (the proof was only finished in May of 2013), but the methods have been thoroughly verified, so we're all convinced that the theorem is true.

The ideas for the proof were first formulated from 1937 to 1939, when two Russian mathematicians with quintessentially Russian-sounding names proved the following:

Vinogradov-Borozdin Theorem Involving a Gigantic Number: *Every odd number greater than $10^{6,846,170}$ can be written as the sum of three primes.*

It is worth noting that this number goes beyond merely "computationally infeasible" and into the realm of just plain stupid. I mean, what the hell are we supposed to do with a number that high? Count up to it? You wouldn't live that long. Neither would your kids. Or grandkids. In fact, the universe would likely be long, long, long gone by the time that the count was finished[5]. So, basically, we've got a theorem involving numbers that aren't even imaginable. You'd have to hope that we could do better than that.

That was indeed the hope for mathematics, and number theorists worked to chip away at this number for the next century or so. The final blow came in 2013, when Harald Helfgott introduced some slick new methods of evaluation that got the bound down to about 10^{30} (i.e. he showed every odd number greater than 10^{30} can be written as the sum of three primes); from there, computers could check all of the values below, and the conjecture was done. Woohoo!

So where does this leave the original Goldbach conjecture? Exactly where it was before: intractable. Most mathematicians believe that the methods used to prove the weak Goldbach won't transfer over to the strong one, which means that we're probably no better off now than we were when we started. But it's still pretty cool to know that every odd number is the sum of three primes.

---

[5] Just for some perspective, $10^{6,846,170}$ is roughly "the number of particles in the universe to the 100,000th power." You can see why it would be tough to count up to that number.

Chapter 36

The Twin Prime Conjecture and Generalizations: Primes Parading in Pairs

36.1 In the Beginning. Well, Not the *Beginning* Beginning. Like, the Beginning of Math, Not The Beginning of Time.

In this chapter, we get to discuss a conjecture that is believed to be one of the oldest unsolved problems in mathematics.

Almost 2,500 years ago, a group of Greek mathematicians began to gather regularly for the purpose of exploring the fundamental questions about the geometry of the universe and, if possible, get out of doing housework. "Look," the early mathematicians would say to their spouses. "I'd love to help clean up around the house, but I have to go explore the fundamental questions about the geometry of the universe!" Often, they would add, "Yes, I know that I used that excuse last week, and...yes, yes, the baby, but...the universe! Fundamental questions!"

From these investigations, the early mathematicians discovered a very important theorem:

Fundamental Rule of the Universe: *Spouses do not care about the fundamental rules of the universe when there is a screaming baby in the house.*

CHAPTER 36. TWIN PRIMES: PARADE!

But they also made a number of other observations. In particular, they discovered many of the basic ideas that allow mathematics to exist today, including the proof, the axiom, the parallel line, the coffee break, the mid-afternoon nap, the NSF grant proposal, and, most importantly, the unhealthy obsession with prime numbers.

It is this last discovery that is of import to the current chapter[1], for it was back in these days that mathematicians first began to look to see whether there were patterns or tendencies amongst the prime numbers. Although many of the patterns they discovered have been lost to history, either because they were obvious ("All prime numbers are prime!") or not particularly mathematical ("Seriously, spouses really, really don't care about prime numbers when there's a screaming baby in the house."), a few of them turned out to be interesting and surprising, and mathematicians quickly set about trying to determine which of these observations could be proven. Several of these could indeed be proven and became the basis of number theory (like the fact that there are infinitely many prime numbers or that no number can have two different prime factorizations), but there were a couple of tantalizingly simple problems that nevertheless confounded the Greeks and, in fact, continue to confound us today.

36.2 Yo, 'Clid!

The most influential of these ancient Greek prime number enthusiasts was Euclid, a famous mathematician who only went by one name because he harbored dreams of someday becoming a rapper[2]. While there were undoubtedly many great mathematicians in ancient Greece, it is no exaggeration to say that Euclid's star likely outshone those of all of his colleagues combined; of the fifty most important observations made by ancient Greek mathematicians, about 30 of them were discovered by Euclid, and the remaining 20 are mistakenly attributed to Euclid because no one remembers any other Greek mathematicians[3]. Euclid found new ideas and theorems in many different

---

[1]And every other chapter in this book.

[2]Sadly, Euclid's rap dreams were dashed when he realized that he had been born 2,200 years before the invention of the microphone. Also, his first album was *horrible*.

[3]Actually, that's not entirely true. One of the observations is falsely attributed to Pythagoras. You might have heard of it? It's called the "Pythagorean Theorem" because

CHAPTER 36. TWIN PRIMES: PARADE! 180

areas of math and philosophy (in fact, geometry has a whole branch known as "Euclidean Geometry"), but in this chapter, we'll limit ourselves to one of his number theory discoveries.

While taking a break from writing horrendous rap lyrics one day, Euclid noticed that prime numbers often show up in pairs that are two apart, like 3 and 5, 5 and 7, 11 and 13, 17 and 19, 29 and 31, 41 and 43, 59 and 61, 71 and 73, and a whole bunch of others. Not all primes show up this way, of course (for example, 37 doesn't show up in one of these pairs, since neither 35 nor 39 is prime), but there seem to be many examples of numbers that do. Euclid wondered how often this was the case; specifically, he wondered if there were infinitely many such pairs or not.

This inspired Euclid to rap the following question[4]:

Euclid's Question: *Are there infinitely many pairs of primes that differ by two?*

Euclid took this to the marketing department, and they decided that "Euclid's Question" was an unacceptable name for a conjecture like this, especially since mathematics already had Euclidean Geometry, Euclid's Division Lemma, the Euclidean Algorithm, Euclid's Guides to Fishing, and the Euclid Dance. They decided to come up with a "snappy" nickname, calling these pairs of primes "Twin Primes" and changing the above to the "Twin Prime Conjecture[5]:"

Pythagoras stole it from the Egyptians.

[4]The actual rap lyrics that he wrote were fortunately destroyed when the Library of Alexandria burned down.

[5]I need to put this in somewhere, so I suppose that this is as good a place as any, but....do you want to hear the nerdiest joke ever? No? Well, too bad - I have to tell it because I just made reference to that "Jenny" (867-5309) song:

"What was Jenny's twin sister's phone number?"
"867-5311"

If you're a human, I'll assume you didn't get the joke, and I'll explain. It turns out, oddly enough, that 8675309 is actually a prime number. Weirder still, it is actually one of a pair of twin primes; the number 8675311 is also prime. So....yeah. That is one nerdy joke.

CHAPTER 36. TWIN PRIMES: PARADE!

If you took all of the famous Greek mathematicians and put them in a list, you would have Euclid.

Twin Prime Conjecture: *There are infinitely many "Twin Primes" (pairs of primes that differ by two).*

The marketers' instincts turned out to be correct. Armed with this nickname, the conjecture quickly became a smash hit, shooting up to the top of the Conjecture Billboard Charts and staying at #1 for over 10,000 weeks before finally being overtaken in 1982 by Michael Jackson's "Thriller."

36.3 Alphonse de Polignac Rides in to Save the Day

Despite attempts by countless mathematicians, this conjecture went unproven for over 2,000 years before Alphonse de Polignac decided in 1849 that it wasn't nearly hard enough and should be made harder. In particular, he realized that, while Euclid had focused on primes that differed by two, he could just as easily have picked another number. For instance, if Euclid had been more fond of the number four, he would have noticed that there are a lot of primes that differ by four: 3 and 7, 7 and 11, 13 and 17, etc. He could also have said the same thing about primes that are six apart. Or eight. Or ten. Not so much eleven. Twelve would work, though. Also 14. 15.6 wouldn't even make sense, but if we rounded off to 16, we might be okay.

What do all of these numbers have in common? You guessed it:

Polignac's Conjecture: *Let k be an even number. Then there are infinitely many pairs of primes that differ by k.*

Some of these pairs have special names. Pairs of primes that are two apart are called "Twin Primes." Pairs that differ by four are "Cousin Primes." Pairs that have difference 8 are called "Co-Worker" primes, while pairs of difference 16 are called "Two People that Saw Each Other on the Street But Haven't Really Talked to Each Other But Wouldn't Be Opposed To It Primes"[6].

Anyway, the importance of de Polignac's work is obvious. Instead of just one impossible question, we have many impossible questions!

36.4 Why Stop There?

It was at this point that all hell broke loose as mathematicians began to realize that there were lots and lots and lots of similar questions that one could ask, and they were all really easy to state. Here are some simple examples:

- *Are there infinitely many numbers n for which n^2+1 is prime?* This is called the n^2+1 conjecture for rather obvious reasons. For example, 4^2+1, which is sometimes written as "17", is a prime number. So is 6^2+1, or 37. 8^2+1 isn't prime, but 10^2+1 (i.e. 101) certainly is. It's believed that there are infinitely many of these n^2+1's that give a prime number.

- *Are there infinitely many prime numbers p for which $2p+1$ is also a prime?* These pairs of primes are known as Germain primes, named for the extraordinary 19th century mathematician Sophie Germain. For example, if you took 23 (which is prime), multiplied by 2 and then added 1, you'd get 47, which is another prime. The same thing would work if you took 29 (prime), doubled it, and added one to get 59 (also prime). There seem to be a lot of these sorts of pairs; in fact we've already found over 20 million pairs of

---

[6]You might think I made the last two up, and that's because I did.

Actually, the only ones that have names are Twin Primes (2 apart), Cousin Primes (4 apart), and, oddly, "Sexy Primes" (6 apart). And no, I didn't make that last one up. I swear.

Germain primes, so it seems entirely plausible that there are infinitely many of these as well.

- Are there infinitely many primes p for which p + 2 and 2p + 1 are both prime? Presumably, this is what it would look like if the Twin Prime Conjecture and the Germain Prime Conjecture had a baby. Incidentally, I've never seen this conjecture attributed to anyone, so we'll call it Wright's Conjecture for now and let historians sort out the details later.

Basically, most equations or sets of equations that you can come up with will probably have infinitely many prime numbers. Do you like the polynomial $x^3 + 3x + 2$? That one's probably prime for infinitely many x's. How about $x^{21} + 3$? Sure, why not. What about the pair $x^{12} + 1$ and $x^4 + 3x + 7$? Yep, those are probably both prime for infinitely many x. You can pretty much go around picking any old polynomials and make a conjecture that they give prime values infinitely often. Go ahead - pick one and name a conjecture after yourself. You can even set up your own Wikipedia page[7]! Don't tell them that I sent you[8].

---

[7]HAH! Yeah, right. Seriously, if you post a Wikipedia page stating your new conjecture about primes of the form $x^{17} + 3x^5 + 4x^2 + 7x + 1$, you'd better take a screenshot really quick, because someone will be taking that page down shortly. Wikipedia is apparently populated by people who do nothing but read Wikipedia all day.

[8]Incidentally, there's actually a sort of meta-conjecture which basically states that any equation or set of equations that you can come up with will have infinitely many primes. It's called the Schinzel Hypothesis or Schinzel-Bouniakowsky Hypothesis. That's right, we now have meta-conjectures in math.

Chapter 37

"Perfect" Numbers: Numbers That are Way, Way Too Full of Themselves

In the last chapter, I mentioned that despite Euclid's many great contributions to mathematics, we would only be discussing one of Euclid's discoveries here - namely, the twin prime conjecture.

Well, I lied. Deal with it.

It turns out that there's another problem of his that we'd like to discuss that, despite the intervening 2,000 years, is still very much an open problem: the problem of the perfect number.

Let me just start out by saying that if I had been in charge of determining how we picked which numbers would be called perfect, I'm pretty sure that I would have never chosen anything like what the Greeks came up with. I mean, one of the examples of their so-called perfect numbers was 28. 28!? 28 is a pretty good number and all, but I certainly wouldn't go so far as to call it perfect. For one thing, it's not even prime!

37.1 Euclidilocks and the Three Factorizations

Anyway, I suppose that at some point, I should explain what this "perfect number" verbiage is supposed to describe.

Let's start out by picking a number. We'll pick 21 because 21 is consid-

CHAPTER 37. PERFECT NUMBERS? NOT ACCORDING TO ME

ered the perfect number for those looking to buy alcohol[1]. Like every other number, 21 has factors, i.e. numbers that divide 21. Just for fun, I'll list them here:

Factors of 21 (besides 21): 1, 3, 7

What's even more fun, of course, is adding those factors together for no apparent reason:

Factors of 21 (besides 21) added together: 1+3+7=11

Man, that was cool.

Because that was so exciting, let's do it again with another number. We'll pick 42 because it's twice as big as 21:[2]

Factors of 42 (besides 42): 1, 2, 3, 6, 7, 14, 21

Factors of 42 (besides 42) Added Together: 1+2+3+6+7+14+21=54

While that was indeed exciting, I suppose that some explanation of why I bothered to do all of this stuff is warranted.

In the first step, when we wrote out all of the factors of 21 and then added them together, we got 11. As some of you may know, 11 is smaller than 21. What this tells us is that 21 doesn't have a lot of factors, and what few factors it does have are pretty small relative to 21. A number where the sum of the factors is less than the number itself (i.e. 11 is less than 21) is considered to be stingy with factors and is therefore called a *cheap son-of-a-gun*.

Actually, wait. I think that might be a typo. A number like this is called a *deficient number*.

By contrast, when we added the factors of 42, we got 54. 54, of course, is bigger than 42. That means that 42 has a surprisingly large number of factors. A number like this (where the sum of the factors is greater than the original number) is not stingy with the factors and thus is said to be *making it rain*, or, alternatively, an *abundant number*.

---

[1] I feel like this joke might need some sort of metric conversion for those reading this outside of the United States.

[2] And also because it takes care of the obligatory *Hitchhiker's Guide* reference.

CHAPTER 37. PERFECT NUMBERS? NOT ACCORDING TO ME

This led Euclid to ask the following question:

Question Which Probably Follows From the Previous Paragraph: *Are there any numbers that are neither abundant nor cheap sons of guns? In other words, are there any numbers for which*

$$Sum\ of\ factors\ of\ n = n?$$

Before I get to the name, I should note that Ancient Greece was the sort of place where people actually worshipped numbers. Like, for real. There were cults where people actually thought it was bad luck to sit on measuring cups because doing so showed disrespect to the numbers on the cup. In fact, Pythagoras became so famous for running a cult that a.) he has a theorem named after him despite the fact that he stole it from the Egyptians and b.) there's a (possibly apocryphal, but possibly not) (but nevertheless believable) story that Pythagoras executed one of his disciples because the disciple proved that $\sqrt{2}$ can't be written as a fraction, thereby destroying the Pythagoreans' belief that everything could be written in terms of fractions[3]. Because of this, Euclid knew that if he gave his concept a cool name, people would randomly start to worship it, and he could subsist on their donations and never have to work again.

Now, about the name. Much like Goldilocks, who had to deal with items that were either too much or not enough before finally finding the perfect bed in which to get attacked and eaten by three angry bears, Euclid realized that he was looking for numbers that somehow split the difference between too much and not enough divisibility and thus decided to call his new numbers *perfect numbers*. In other words,

Definition: *A number n that equals the sum of its divisors and therefore satisfies the bolded question above will be called a <u>perfect number</u>.*

In other words (for those who like equations), we would say that a number n is a perfect number if

$$Sum\ of\ factors\ of\ n = n.$$

---

[3] I wanted to note that there's some ambiguity as to whether this story is true or false because otherwise the reader might think it is simply false like every other biographical story in this book.

CHAPTER 37. PERFECT NUMBERS? NOT ACCORDING TO ME

37.2 Actually Finding Perfect Numbers

Knowing that he was now financially set for life, Euclid set about the task of actually looking for such numbers. He found the first one fairly easily:

Factors of 6: 1, 2, 3
Factors of 6 Added Together: 1+2+3=6
Conclusion: 6 is a perfect number. All hail 6!

The next one wasn't much harder:

Factors of 28: 1, 2, 4, 7, 14
Factors of 28 Added Together: $1 + 2 + 4 + 7 + 14 = 28$
Conclusion: 28 is a perfect number, too.

After that, though, there's a little bit of a gap...

Factors of 496: 1, 2, 4, 8, 16, 31, 62, 124, 248
Factors of 496 Added Together: $1+2+4+8+16+31+62+124+248 = 496$
Conclusion: 496 is equally deserving of our worship. All tithe checks should be payable to Euclid.

...which is then followed by a larger gap:

Factors of 8128: 1, 2, 4, 8, 16, 32, 64, 127, 254, 508, 1016, 2032, 4064
Factors of 8128 Added Together: 8128 (what am I, a calculator? Add them yourself if you don't believe me.)
Conclusion: 8128 - yeah, you know the drill.

And...this was the point where Euclid was done. He found his four perfect numbers, put them on his fliers, announced them to the world, and retired to a tax-sheltered island in the Carribean where no one would ever be able to track him down[4].

---

[4]I should note that in reference to some of the later perfect numbers here, when I say "Euclid discovered", what I mean is "someone discovered, and then mathematicians mistakenly attributed to Euclid." For instance, the perfect number 8128 is actually believed to have been by someone named Nichomachus, who lived about 300 years after Euclid. Do you care about Nichomachus? Me neither. Let's call him "Euclid" and be done with

37.3 Even More Perfect Numbers

While it seems like Euclid was being lazy by stopping where he did, he actually had a very good reason: his calculating technology basically consisted of an abacus, a pen, and a couple barrels of wine, and thus it would have been quite an ordeal to try to compute the divisors of numbers over, say, 10,000. As a result, the discovery of the next perfect number waited for another 1500 years until computation methods improved and the wine ran out.

In 1456, things had finally improved to the point where people were able to think about new perfect numbers again, and a couple of mathematicians even went so far as to try to find the next one. One of these mathematicians actually succeeded, but someone forgot to write down his name, so now we have no idea who it was. All we know is that whoever discovered it lived in Florence somewhere in the late 1450s. Also, he had a lot of time on his hands, because he found a big one:

Factors of 33,550,336: This is left as an exercise to the reader because I'm sure as hell not doing it.

Factors of 33,550,336 Added Together: Probably 33550336, since Wikipedia lists it as a perfect number.

Having established that there was more work to be done, mathematicians decided to try to find a few more. The next breakthrough was made by Pietro Cataldi, who came up with a brilliant formula that gave six perfect numbers, three of which were wrong, and another of which he speculated was a perfect number but never actually bothered to check. So, um, that's progress:

Factors of 8589869056 and 137438691328: Add up to themselves.

Factors of 35184367894528, 144115187807420416, and 9444732965670570950656: Don't add up to themselves, but Cataldi figured no one would actually check, so he didn't bother to figure that out for himself.

Factors of 2305843008139952128: Actually do add up to 2305843008139952128, but Cataldi got bored halfway through the computa-

it.

CHAPTER 37. PERFECT NUMBERS? NOT ACCORDING TO ME

tion and never actually figured out whether this was the case. Euler eventually proved this to be true about 200 years later.

Nowadays, we actually know that there are as many as 48 perfect numbers. As number theorists, we figure that nothing interesting comes in groups of 48, so we've decided, with no evidence whatsoever, that there must be infinitely many:

Perfect Number Conjecture: *There are infinitely many perfect numbers.*
Proof: *???*

37.4 How Did They Find These Numbers, Anyway

37.4.1 Gettin' Gimpy With It: Mersenne Primes

As you may suspect, it turns out that the search for perfect numbers is a little bit easier than "Take every number, figure out all of its factors, and then add them up." Indeed, there is a very clever trick that makes the search go much, much faster. Of course, in this case "much, much faster" is still pretty glacial, given that we as a species have averaged about one new perfect number every fifty years or so.

Anyway, recall that in one of the footnotes, I said that the fourth of the perfect numbers was discovered by Nichomachus, or, as we called him earlier, "Euclid." Well, Nichomachus, had an ace up his sleeve, which is weird when you consider that he lived before the invention of playing cards, and also that he was an ancient Greek and therefore wore a sleeveless toga. He discovered that it was much easier to find perfect numbers if you made use of a type of prime number that he named a *Mersenne Prime* in honor of Marin Mersenne, a monk who lived sixteen hundred years after Nichomachus died[5].

To understand these primes, I pose the following observation:

Interesting Observation: *Consider the expression $2^n - 1$. Or don't. Fine. Just sit there. See if I care.*

---

[5]OK, I'll admit it: every time I hear the name Nichomachus, I think of the "Nicoderm CQ" patch. Somehow, in my head, perfect numbers and quitting smoking are forever linked. I'm not sure what to do about this.

CHAPTER 37. PERFECT NUMBERS? NOT ACCORDING TO ME

Oops, sorry - I got a bit defensive there. Let's try that again:

Another Interesting Observation: *Consider the expression $2^n - 1$. Sometimes, if you plug in n, numbers come out. Like, if I plugged in $n = 3$, I'd get $2^3 - 1 = 7$. Or if I plugged in $n = 4$, I'd get $2^4 - 1 = 15$. Or maybe even $n = 7$, which would give me $2^7 - 1 = 127$.*

Note that sometimes, these numbers that come out are prime. And sometimes, they aren't.

The question, then, is this:

The Question: *For which n is $2^n - 1$ prime?*

We know that $n = 3$ works (since $2^3 - 1 = 7$ is prime), and we know that $n = 4$ doesn't work (since 15 isn't prime), and some of you may have known that $n = 7$ also works (since 127 turns out to be prime). What's the pattern?

Answer: *In order for $2^n - 1$ to be prime, n must itself be a prime number. And other things have to happen, too. Like n not being 11. Or 23. Or 29, 37, 41, 47, 53, or a whole list of other primes that happen to not work.*

In other words, there's some other criterion besides n being prime that we haven't figured out yet. So...we don't know.

To put this in other words, for $2^n - 1$ to be a prime, n has to be a prime number. That said, just knowing that n is prime is not enough - n being prime is just a ticket to get into the "is $2^n - 1$ prime?" game, but it's not enough to actually guarantee that you win the game. This is important, because, as Charlie Sheen pointed out, it's all about winning. And having tiger blood and Adonis DNA.

Hold on, I think I got sidetracked there. What was I saying? Oh, that's right, this gets us to the point where we actually get to name something:

Definition: *A **Mersenne Prime** is a prime number of the form $2^n - 1$. Of course, n has to be prime.*

Incidentally, you know how sometimes you'll read that someone found a new largest known prime number? If you see one of those articles, there's about a 100% chance that the number was a Mersenne Prime. In fact, there's actually a project dedicated to finding larger and larger Mersenne Primes called, oddly enough, GIMPS (the Great Internet Mersenne Prime Search). You can download some software onto your computer, and your CPU can start grinding through various options for Mersenne Primes when it isn't doing anything else (like when your computer is asleep).

You might be wondering why people bother looking for the new largest known prime number, given that there are infinitely many and, hence, the supply is literally inexhaustible. Well, it turns out that there's a very good reason: there's a company (called EFF) that randomly pays out money to people who can find large primes. In fact, in 2009, GIMPS walked home with $100,000 for finding a 12 million digit prime. That's a pretty good chunk of change for something that your computer can do while you're off drinking.

37.4.2 Practicing Mersenne Primes Make Perfect

While Mersenne Primes are pretty interesting beings on their own, they're also useful for finding perfect numbers (which makes sense, given that I put them in the "Perfect Numbers" chapter). The reason that they're so useful is because every Mersenne prime corresponds to a perfect number, and every even perfect number corresponds to a Mersenne Prime:

Amazing Observation: *If $2^n - 1$ is a Mersenne Prime then $2^{n-1}(2^n - 1)$ is a perfect number. Likewise, any even perfect number can be written as $2^{n-1}(2^n - 1)$ for some Mersenne Prime $2^n - 1$.*

For instance, we saw that $2^3 - 1$ was prime, since it was 7. The Amazing Observation then tells us, without any further work whatsoever on our part, that $2^2(2^3 - 1)$ must therefore be a perfect number. (It's actually 28, which is one of the perfect numbers Euclid found.)

As you might imagine, this observation certainly made finding perfect numbers quite a bit easier. After all, for Cataldi or Euler or Nicoderm CQ to find new perfect numbers, they just had to look for a Mersenne prime (which was a lot easier, since they could just plug in values for n until they got one that works), and, presto, they had a perfect number by the observation above! This trick enabled these three mathematicians to expand the list of perfect numbers to as many as ten before they had to give up and call in the computers for help[6].

37.5 Still Looking

It is amazing to think that even though the concept of a perfect number was discovered over two thousand years ago, many of the simplest questions about them persist to this day. Are there infinitely many? Are any of them odd? Can EFF give me $100,000 for looking for one, too? Is it really possible that Goldilocks outran all three of the bears, especially since a.) Papa Bear was probably way bigger than she was and b.) she was fast asleep when the bears found her? One can't say. We can only hope that someday, mathematics will help us to answer these baffling riddles once and for all.

---

[6]Note that I stipulated that the amazing observation is true for any *even* number that happens to be a perfect number. That's because we don't know anything about odd numbers that are perfect numbers. We don't know if they exist, or, if they did, what they'd look like. All we know is that if these mythical creatures did exist, they'd probably be really big. Basically, odd perfect numbers are like the Loch Ness Monsters of the mathematics world.

Chapter 38

Epilogue

Anyway, these are the problems that we use in order to suck people into number theory. These conjectures sure look easy, don't they? Just go ahead and sit down and start working one of them, and, surely, something will come out....

Chapter 39

Acknowledgements

A little biographical story: when I was two years old, my mom used to walk me down the street and I would excitedly announce the mailbox numbers. There was no particular reason for me to do this; two year-olds do whatever they're interested in, and little Tommy happened to be fascinated by numbers. All of my mom's friends used to call me The Count. If anyone ever wants to know when I started getting into math....it's been a lifelong thing.

Anyway, I share this as a posthumous way to say thanks to Mom and Dad for encouraging me to do what I was interested in, even beyond the point where they ceased to understand what I was doing. Thanks also Cheryl, Jamie, Jen, Emily, and Aleyna for putting up with me and my math habit. Thanks to Doc Harrison for recognizing my passion for math at a very early age (6 years old) and nurturing my talent so that I would one day become a mathematician.

It took me a little while to figure out what this book was going to be. The idea for this sort of friendly introduction to number theory actually came to me when I was defending my Ph.D. dissertation in 2009. Four of the five committee members on my dissertation defense committee were mathematicians, and the fifth was lost - he was a neuroscientist with no background in number theory who was largely on my committee as a personal favor. So I wrote "A Friendly Introduction to My Thesis" in an attempt to help him out. I'm not sure if my attempt actually succeeded, but the Friendly Introduction helped me find the writing style that I employ in this book. So thanks to Dr. Stewart Hendry and Michelle Brown for helping me discover my inner writer.

A huge thank you is in order for Jeremy Henkel, who undertook the

Herculean task of reading this entire book from cover to cover and giving copious edits and suggestions. Thanks also to others who made suggestions to make this book better, especially Vickie Kearn, Lola Thompson, Matt Dingledine, Matt Cathey, and everyone else I'm forgetting. Thank you to Wofford writers' group for keeping me on task. Thanks to all my students who encouraged me to write this book, especially those in my number theory, appreciation/patterns, and proofs classes who acted as beta testers and proofreaders for the book. Special thanks to Daniel Goetz for helping me come up with a working version of the title.

Thanks also to Michelle Griggs for a lovely cover.

There's not enough space here to thank all of the teachers who made this book what it is, so let me send my biggest thank you to all of my teachers at Easton Public Schools and Bowdoin College, all of my teachers and colleagues at Johns Hopkins, and all of my colleagues at Lawrence University and Wofford College.

Chapter 40

Photo credits

Just as an upfront comment, only one of the photos I used required any consent from any of the people involved. Most of them were on Creative Commons or in the public domain, which means that a.) I didn't have to pay for them, b.) I have no rights to them, and c.) no one in the photo, near the photo, taking the photo, described by the photo, affiliated with the photo, near anyone affiliated with someone described in the photo, or named "Photo" gives any endorsement to my use of the photos or of this book. Most of them probably don't know of this books' existence. The goal of this section is to make it so that their lawyers don't have to know about my book's existence, either. To wit, here are some words about photos:

Page 16 (Hilbert): This picture is in the public domain. It appears on Wiki Commons.

Page 23 (Riemann): This picture is in the public domain. It appears on Wiki Commons.

Page 29 (Gauss): This picture is in the public domain. It appears on Wiki Commons.

Page 31 (Bruce Lee): This picture is taken from a stamp issued by the Republic of Tajikistan and hence is not subject to any copyright protections.

Page 48 (wrestling photo with Andre the Giant) (2004): This photo, uploaded to the Creative Commons/Flickr by "Ethan" and later posted to

CHAPTER 40. PHOTO CREDITS

Wikipedia's WikiCommons by "Techarrow," is subject to a ShareAlike License. Neither the originator of the photo nor Creative Commons has endorsed my use.

Page 63 (broken computer) (2008): This photo, uploaded to Wikipedia's WikiCommons by user "M2545," is subject to a ShareAlike 2.0 License. Neither the originator of the photo nor Creative Commons has endorsed my use.

Page 116 and 131 (Cheech and Chong) (1977): This publicity photo from American Bandstand is in the public domain, since it appeared before 1978 and lacks proper copyright markings. The use of this photo does not imply authorization by Cheech, Chong, American Bandstand, ABC, or any partners thereof.

Page 95 (Eazy E/Euler): The picture of Eazy E is used with the permission of Corbis Images. All rights are reserved by Corbis. The picture of Euler is public domain and appears on Wikipedia.

Page 108 (Bryan Birch): The photo of Birch was place on Wikipedia's Creative Commons Wikipedia by William Stein. It is covered by Creative Commons Attribution 3.0 and requires proper attribution. The use of this photo does not imply endorsement by Birch, William Stein, or anyone else.

Page 108 (Sir Peter Swinnerton-Dyer): The photo of Swinnerton-Dyer was placed on Wikipedia's Creative Commons Wikipedia by the Oberwolfach Photo Collection. It is covered by Creative Commons Attribution 2.0 and is subject to a ShareAlike License. The use of this photo does not imply endorsement by Swinnerton-Dyer or Oberwolfach, or any organizations related to either.

Page 145 (Erdos): This picture was uploaded to Wikipedia's WikiCommons by user "PaultheOctopus." The original picture had Erdős and a young Terence Tao; the picture has been altered to show only Erdős. It is subject to a Sharealike 2.0 License.

Page 169 (Stargell): This photo, taken and uploaded to Wikipedia's WikiCommons by Sally Lindsay, is subject to a ShareAlike 2.0 License. Neither the originator of the photo nor Creative Commons has endorsed my use.

Chapter 40. Photo Credits

Page 173 (Goldbach): This picture is in the public domain. It appears on Wiki Commons.

Page 181 (Euclid): This picture is in the public domain. It appears on Wiki Commons.

Made in the USA
Columbia, SC
22 January 2020